Variability and Management of Large Marine Ecosystems

AAAS Selected Symposia Series

Published by Westview Press, Inc.
5500 Central Avenue, Boulder, Colorado

for the

American Association for the Advancement of Science
1333 H Street, N.W., Washington, D.C.

Variability and Management of Large Marine Ecosystems

*Edited by Kenneth Sherman
and Lewis M. Alexander*

AAAS Selected Symposium **99**

AAAS Selected Symposia Series

This book is based on a symposium that was held at the 1984 AAAS Annual Meeting in New York City, New York, May 24-29. The symposium was sponsored by AAAS Section G (Biological Sciences) and Section W (Atmospheric and Hydrospheric Sciences).

This Westview softcover edition was manufactured on our own premises using equipment and methods that allow us to keep even specialized books in stock. It is printed on acid-free paper and bound in softcovers that carry the highest rating of the National Association of State Textbook Administrators, in consultation with the Association of American Publishers and the Book Manufacturers' Institute.

Published in 1986 in the United States of America by Westview Press, Inc.; Frederick A. Praeger, Publisher; 5500 Central Avenue, Boulder, Colorado 80301

Library of Congress Cataloging in Publication Data
Main entry under title:
Variability and management of large marine ecosystems.
    (AAAS selected symposia series; #99)
    1. Marine ecology. 2. Marine resources--Management. I. Sherman, Kenneth, 1932-   . II. Alexander, Lewis M., 1921-   . III. Series: AAAS selected symposium; #99.
QH541.5.S3V35   1985   333.95'2   85-26489
ISBN 0-8133-7145-7

Composition for this book was provided by the editors.
This book was produced without formal editing by the publisher.

Printed and bound in the United States of America

 The paper used in this publication meets the minimum requirements of the American National Standard for Permanence of Paper for Printed Library Materials Z39.48-1984.

6   5   4   3   2   1

About the Book

Large marine ecosystems (LMEs) are being subjected to increasing stress from industrial and urban wastes, aerosol contaminants, and heavy exploitation of renewable resources. Recent studies suggest the population structure of the LMEs can be altered by these factors with a resulting negative economic impact. Ecosystem perturbations have been documented from the Bering Sea to the Antarctic and the Gulf of Thailand to the El Niño region off the Peruvian coast. This book is a state-of-the-art review of effective means for measuring changes in populations and productivity, physical-chemical environments, and management options for LMEs. For the first time, large marine ecosystems are treated holistically as regional management units bringing together the all too often fragmented efforts to optimize ocean resources. Strategies for measuring the natural variability are examined against a background of anthropogenically induced pollution and overexploitation. It is becoming increasingly obvious that a multidisciplinary approach involving science, law, economics, and government is necessary to address the biological, physical, judicial, and socioeconomic problems that need to be dealt with if management is to be successful.

Dr. Kenneth Sherman is director of the Narragansett Laboratory, Northeast Fisheries Center, National Marine Fisheries Service, National Oceanic and Atmospheric Administration (NOAA). His coeditor, *Dr. Lewis M. Alexander,* is director of the Center for Ocean Management Studies, University of Rhode Island.

About the Series

The *AAAS Selected Symposia Series* was begun in 1977 to provide a means for more permanently recording and more widely disseminating some of the valuable material which is discussed at the AAAS Annual National Meetings. The volumes in this *Series* are based on symposia held at the Meetings which address topics of current and continuing significance, both within and among the sciences, and in the areas in which science and technology have an impact on public policy. The *Series* format is designed to provide for rapid dissemination of information, so papers are reproduced directly from camera-ready copy. The papers are organized and edited by the symposium arrangers who then become the editors of the various volumes. Most papers published in the *Series* are original contributions which have not been previously published, although in some cases additional papers from other sources have been added by an editor to provide a more comprehensive view of a particular topic. Symposia may be reports of new research or reviews of established work, particularly work of an interdisciplinary nature, since the AAAS Annual Meetings typically embrace the full range of the sciences and their societal implications.

WILLIAM D. CAREY
Executive Officer
American Association for
the Advancement of Science

Contents

8   RESULTS OF RECENT TIME-SERIES OBSERVATIONS
    FOR MONITORING TRENDS IN LARGE MARINE
    ECOSYSTEMS WITH A FOCUS ON THE NORTH
    SEA--Niels Daan

9   COMPARISON OF CONTINUOUS MEASUREMENTS AND POINT
    SAMPLING STRATEGIES FOR MEASURING CHANGES IN
    LARGE MARINE ECOSYSTEMS--Alex W. Herman

Tables

Figures

xix

About the Editors and Authors

Kenneth Sherman is director of the Narragansett Laboratory at the Northeast Fisheries Center of the National Marine Fisheries Service, National Oceanic and Atmospheric Administration (NOAA), and an adjunct professor of oceanography of the Graduate School of Oceanography, University of Rhode Island. His specialty is the recruitment process within fish populations, and he is the author of numerous papers on fisheries ecology. In addition to his research on the northeast continental shelf of the United States, he serves as chairman of the Biological Oceanography Committee of the International Council for the Exploration of the Sea and is chief scientist of the Antarctic Program of the National Marine Fisheries Service.

Lewis M. Alexander is director of the Center for Ocean Management Studies and professor of geography at the University of Rhode Island. He has served as geographer of the U.S. Department of State and is a member of the United States Delegation to the Law of the Sea. He is a pioneer in the development and practice of a course of study for graduates and undergraduates in marine policy and he serves on the editorial board of several scholarly journals.

Andrew Bakun is chief of the Pacific Fisheries Environmental Group at the Southwest Fisheries Center of the National Marine Fisheries Service, NOAA, in Monterey, California. His specialty is linkages between physical and biological ocean processes, with particular reference to exploited fish stocks. He is currently chairman of the Guiding Group of Experts for the Program of Ocean Science in Relation to Living Resources, a new scientific program cosponsored by the Intergovernmental Oceanographic Commission and the Food and Agricultural Organization of the United Nations.

John R. Beddington is director of the Marine Resources Assessment Group in the Centre for Environmental Technology at Imperial College of Science and Technology in London, England. He has been active in the development and use of large marine ecosystem models. In addition to serving as scientific advisor to the International Institute for Environment and Development and also the International Union for Conservation of Nature and Natural Resources, he is a representative of the United Kingdom to the Commission for the Conservation of Antarctic Marine Living Resources.

Martin H. Belsky is associate professor at the University of Florida College of Law. He has written numerous articles on U.S. and international environmental and ocean law, coastal zone management, and domestic and international fisheries issues. He is a former chief counsel for the U.S. House of Representatives Select Committee on the Outer Continental Shelf and also a former assistant administrator for the National Oceanic and Atmospheric Administration.

William Y. Brown, director of marine affairs for Waste Management, Inc. in Washington, D.C., specializes in environmental law and policy. In addition to his congressional and agency testimony on the Arctic and Antarctic, he has written numerous articles on ecosystem conservation.

John Byrne is president of Oregon State University, where he was formerly the dean of the School of Oceanography. He has also served as administrator of the National Oceanic and Atmospheric Administration. In addition, he served as the United States Commissioner to the International Whaling Commission, where he was instrumental in the enactment of measures taken by the IWC to accelerate the recovery of depleted global whale stocks.

Francis T. Christy, Jr., is senior fishery planning officer at the Food and Agriculture Organization of the United Nations in Rome, Italy. In addition to numerous articles on fisheries economics and management, he has written The Common Wealth in Ocean Fisheries (with Anthony Scott; Johns Hopkins Press, 1962) and Alternative Arrangement for Marine Fisheries (Resources for the Future, 1973).

Niels Daan is professor in the marine department at The Netherlands Institute for Fishery Investigations in Ijmuiden, The Netherlands. His specialty is fisheries biology and he has written numerous articles on species interactions and exploited fish population dynamics.

Robert L. Edwards has a long and distinguished career as a research administrator in the National Marine Fisheries Service. He has served as director of the Northeast Fisheries Center at Woods Hole, assistant director of the National Marine Fisheries Service in Washington, D.C., and special assistant to the administrator for science of the National Oceanic and Atmospheric Administration (NOAA). He has published extensively on multispecies interactions and the impacts of environmental perturbation on fisheries resources, and he is the recipient of numerous awards, including the Department of Commerce Gold Medal for distinguished contributions to marine ecosystems studies.

Alex W. Herman, a nuclear physicist by training, is a research scientist at the Bedford Institute of Oceanography in Dartmouth, Nova Scotia. He has written extensively on marine ecology and biological instrumentation.

Lewis Incze is a research scientist at the College of Ocean and Fishery Sciences at the University of Washington, Seattle. His specialty is larval ecology and recruitment and also the relationships between physical features and patterns of planktonic production in estuarine and marine environments. He has received awards from the National Shellfisheries Association and the American Institute of Fisheries Research Biologists.

Gunnar Kullenberg, professor of physical oceanography at the University of Copenhagen in Denmark, is a member of the Royal Danish Academy of Sciences and Letters. In addition to writing numerous articles related to Baltic and North Sea oceanography, he is the author of The State of the Baltic (Pergamon, 1981) and editor of Pollutant Transfer and Transport in the North Sea, Volumes I and II (CRC Press, Inc., 1982).

Alec D. MacCall is a fishery biologist at the Southwest Fisheries Center of the National Marine Fisheries Service, NOAA, in La Jolla, California. His specialty is fishery stock assessment and the dynamics and management of coastal pelagic fishes. He has served as chairman on numerous committees, including the Anchovy Plan Development Team of the Pacific Fishery Management Council, the Committee on Fisheries and Seabirds of the Pacific Seabird Group, and the Stock Assessment Committee of the California Cooperative Oceanic Fisheries.

Bruce S. Manheim is a scientist in the wildlife program at the Environmental Defense Fund. He specializes in wildlife law and has written articles on the conservation of Antarctic marine living resources. In addition, he is a U.S. delegate to the Convention on the Conservation of Antarctic Marine Living Resources.

Giulio Pontecorvo, professor in the Graduate School of Business at Columbia University, specializes in fishery economics. He has written several books, including Fisheries Conflicts in the North Atlantic: Problems of Jurisdiction and Enforcement (Ballinger, 1974), Law of the Sea: Emerging Regime of the Oceans (Ballinger, 1974), and The New Order of the Oceans: The Advent of a Managed Environment (Columbia University Press, forthcoming).

J. D. Schumacher, an oceanographer at Pacific Marine Environmental Laboratory in Seattle, Washington, specializes in fisheries oceanography and physical processes over continental shelves. His papers have appeared in numerous journals.

R. Tucker Scully is director of the Office of Oceans and Polar Affairs at the U.S. Department of State. He is active in Antarctic Ocean matters and presently serves as head of the U.S. delegation to the Convention on the Conservation of Antarctic Marine Living Resources.

Michael P. Sissenwine is chief of the Fisheries Ecology Division at the National Marine Fisheries Service, NOAA, in Woods Hole, Massachusetts. He has published extensively on fisheries management.

Preface

Large Marine Ecosystems (LMEs) adjacent to land masses are being subjected to increased stress from heavy exploitation of renewable resources and used as disposal areas for industrial and urban wastes and aerosol contaminants. These areas are characterized by unique hydrographic regimes, submarine topography, and trophically-dependent populations. Recent reports indicate that the population structure of LMEs can be altered by natural and anthropogenic changes. Perturbations resulting in dominance shifts among the predator field of LMEs have been reported for the North Sea, Gulf of Thailand, northeast United States coast, the Antarctic, California Current, El Niño region off Peru, the East Bering Sea, and the Baltic. The consequences of these shifts have had far-reaching socioeconomic impact.

The first symposium on the Variability and Management of Large Marine Ecosystems was convened at the 1984 Annual Meeting of the American Association for the Advancement of Science. It provided a forum to review strategies for measuring the natural variability of LMEs against a background of increasing evidence of anthropogenically induced perturbations from over-exploitation and pollution. Each of the invited participants was encouraged to synthesize the growing, but scattered, body of information that would have a bearing on improving means for measuring ecological changes and developing management options for LMEs. The results of these syntheses have been brought together for this volume to increase the awareness among resource management organizations, scientists, lawyers, and students that it is scientifically, technically, and legally feasible to implement holistic conservation and management regimes for LMEs. The economic aspects of LME management are explored with regard to the cost and benefits of alternative management regimes and the prognoses for management leading to optimal yields of living resources within LMEs.

The volume is arranged in three parts. The first deals with the impacts of perturbation on the productivity of renewable resources in LMEs. The second part focuses on measuring variability in LMEs. The volume concludes with a consideration of the legal, economic, and strategic framework for managing LMEs.

The editors are indebted to the participants for their willingness to take time out of busy schedules to provide the expert syntheses and reviews needed to put into focus the scale of events that should be considered for more effective management of the products of natural production in LMEs. We are especially indebted to Jennie Dunnington for her skilled assistance in overseeing the editorial production of the final manuscript. Thanks also are extended to Mattie Walker for her assistance in typing the contributions.

K. Sherman
L. Alexander
Narragansett, Rhode Island
March, 1985

Impact of Perturbations on the Productivity of Renewable Resources in Large Marine Ecosystems

1. Introduction to Parts One and Two: Large Marine Ecosystems as Tractable Entities for Measurement and Management

The movement toward total ecosystem management has been growing slowly within the international community for several decades. This trend began with the deliberations of the International Council for the Exploration of the Sea (ICES) in its first meetings conducted at the turn of the century. The meetings were prompted by the realization that the capacity for the oceans to produce an inexhaustible supply of commercially desirable fish species was finite, and that overfishing could result in serious depletion of the stocks. The first attempts to deal with the management of large marine ecosystems (LMEs) took place in Kristiania, Norway, in 1901 where representatives from Denmark, Germany, Norway, Russia, and the United Kingdom set the course for the establishment of ICES with a series of resolutions directed to the establishment of joint international biological and hydrographic studies to be conducted in the North Atlantic, the North Sea, and the Baltic Sea. During the intervening 84 years, ICES has provided a fertile ground for the development of joint international studies of marine ecosystems. In the process, generations of scientists participating in the work of ICES helped to focus attention on the advantages of coordinated multidisciplinary studies of these and other LMEs. The LMEs are defined as regions with unique hydrographic regimes, submarine topography, and trophically-linked populations. As the trend for management of living resources moves from single species to multispecies assemblages, it becomes increasingly important to encompass entire ecosystems as management units. This action will ensure that measures designed to optimize the natural productivity of target species assemblages will also include considerations for related populations and their environments.

The attempt to deal adequately with conservation and management of living resources in LMEs has evolved from early ICES beginnings to the state described by J. Beddington, who in his contribution to the volume addresses the extent to which changes produced by exploitation of marine ecosystems are reversible. Beddington argues that the most appropriate common ground for moving forward in the management of resource populations in large ecosystems is not with total ecosystem models, nor the single species recovery models exclusively, but rather with an approach that covers the common ground between observation and theory at the multispecies level, based on models focused on ecological phenomena. He concludes that, even now, more effort needs to be directed to mesoscale monitoring of resource populations and their environments to reduce the difficulties in distinguishing between man-induced and natural changes in ecosystem components.

The LMEs are tractable units for total ecosystem management. Several case histories demonstrate that when measurements of populations and their environments are made on the appropriate temporal and spatial scales, the sources of ecosystem change can be identified. G. Kullenberg attributes the source of ecosystem change in the Baltic Sea to man's activities. He provides a comprehensive treatment of the Baltic as a LME that has undergone significant changes during the past 80 years. He argues that the increase in the volume of anoxic water from organic terrigenous sources and increasing levels of primary productivity has shifted conditions in the coastal zones from oligotrophic to eutrophic. As a consequence, oligotrophic fish, including perch and pike, are being replaced by the eutrophic bream and roach fishes. In addition, the overall fish biomass yields have doubled over the past 20 years. Large increases have been observed in the benthic biomass and in the cod stocks in particular. From an ecosystem management perspective, Kullenberg urges that limits are needed on inputs of contaminants and that these limits should be made part of a control and monitoring system to evaluate the results of management activities.

The value of decadal time-series information based on mesoscale population and environmental changes in the California Current LME is apparent in the case study prepared by A. D. MacCall. The lack of a clear understanding of predator-prey relationships among the high biomass species has limited the possibilities for separating the impacts of natural variability from fishery impacts on the pelagic fish components of the ecosystem. The chapter includes a discussion of the abundance trends in other important California Current ecosystem populations including sea birds and pinnipeds. The importance of multispecies interaction is stressed. MacCall concludes that achievement of an

ecosystem management regime for the California Current
ecosystem will require a more consolidated federal-
state management authority than presently exists.

A case study of an ecosystem under stress is
provided by M. P. Sissenwine. He argues that the
decline in fish biomass on the northeast continental
shelf off the United States could have been more
drastic than the 60% decrease from the mid-1960s to the
mid-1970s. The decrease is attributed to extremely
high fishing mortalities on the spawning biomass of the
commercially-important fish species. Sissenwine points
out that while pre-exploitable fish represent only 10%
of the biomass of exploitable fish, their predation
impact ranges between 60% and 90% of their own
production. The other important ecosystem consumers
are identified by Sissenwine as marine mammals, birds,
large predatory fish, and man. Approximately 25% of
ecosystem production is consumed by mammals, birds, and
man. The only other components with significant biomass are the large pelagics estimated as representing
approximately 20% of the remaining biomass. The
changes in biomass are attributed to the impacts of
predation within the tightly bound energetics of the
northeast shelf ecosystem. The impacts of "man-the-
predator" are shown to be far greater than any natural
environmental perturbation in the northeast shelf LME.

The second section of the volume deals with
several aspects of measuring variability within LMEs.
A. Bakun describes the importance of attempts to bring
together generalizations from similar LMEs, albeit
widely scattered geographically, to make better use of
the limited empirical data available. He provides
examples from eastern boundary currents that demonstrate the similarities of upwelling pelagic ecosystems
including the California Current, Canary Current, Peru
Current, and Benguela Current LMEs. The four LMEs
appear to be controlled by similar environmental events
and support similar communities of pelagic species.
Bakun focuses on the importance of understanding reproductive strategies of the fishes that appear to be
adapted to maintaining spawning products within the
more highly productive nearshore upwelling areas and
minimizing offshore transport of eggs and larvae.
Optimal sampling strategies are discussed with respect
to the monitoring of large-scale environmental conditions.

The importance of natural environmental events on
the survival of living resources of the eastern Bering
Sea is discussed by L. Incze and J. D. Schumacher.
They argue that year-class strength of fish within the
eastern Bering Sea LME is influenced by oceanographic
signals. In addition, they have examined the potential
impact of inter- and intraspecific predation on

recruitment failure. This study provides a solid basis from which to proceed to specific hypothesis testing and has resulted in a rather significant commitment of funds for full-scale oceanographic and aerial support for conducting a fishery oceanography experiment (FOX) over the entire eastern Bering Sea during 1985.

The importance of time-series observations separating fisheries-induced variability from natural variability is emphasized in the paper by N. Daan. Based on an extensive 70-yr data set from the North Sea, Daan demonstrates that ongoing monitoring programs within the North Sea are adequate for detecting general trends, but they are not sufficient for understanding the key processes controlling recruitment success. The addition of a systematic collection and analyses of 40,000 fish stomachs allowed for an unprecedented examination of the species composition and average number of prey consumed by mid-size fish predators within the North Sea ecosystem. Daan presents convincing evidence that a good bit of uncertainty exists in the use of age structured catch data to estimate population sizes of cod, haddock, whiting, and herring in the North Sea LME. The combined approach of fishery-independent surveys and environmental measurements are stressed by Daan as a critically important strategy for measuring variability within LMEs, particularly in recognition of the deterioration of the reliability of catch statistics resulting from abuses in the reporting of catches under the total allowable catch management system.

Advanced methods for measuring chlorophyll, primary productivity, and copepod variability in LMEs are discussed by A. W. Herman. His studies are based on a comparison between classical underway net sampling and more advanced point sampling with LMEs of the Eastern Canadian Arctic, the Scotian Shelf, the Peruvian Shelf and the Eastern Tropical Pacific ecosystems. The study demonstrates clearly the importance of considering vertical structure in any sampling design for relating plankton to hydrographic features. The unique comparison among tropical, temperate, and arctic LMEs underscores the differences in sampling strategies required to reduce variability and achieve accurate representation of the planktonic components of LMEs.

The approach underway in the United States for systematically measuring living resources within the Exclusive Economic Zone (EEZ) is given in the contribution by K. Sherman. Much of the discussion focuses on the utility of combined hydrographic and ichthyoplankton surveys conducted on mesoscale grids of 20 to 100 km at frequencies of 2 to 12 times per year. The results of these marine ecosystem studies are discussed for several of the LMEs under investigation with a focus on the success achieved in measuring shifts in the dominance of fish species within LMEs, including sardines and anchovies in the California Current Ecosystem, the importance of pollock biomass in the Eastern Bering Sea Ecosystem and the apparent replacement of herring by sand eel in the Northeast Continental Shelf Ecosystem.

Collectively, these studies argue for the need to include multidisciplinary biological and environmental mesoscale studies aimed at measuring and forecasting changes in the dominant exploitable species of LMEs in support of multispecies conservation and management regimes. This need is not significantly different from the principles articulated by the founders of ICES at the turn of the century.

John R. Beddington

2. Shifts in Resource Populations in Large Marine Ecosystems

ABSTRACT

This paper addresses the problem of the extent to which changes in marine communities produced by exploitation are reversible. In the context of ecological theory this raises the possibility that appropriate models of marine communities possess alternative stable states. The implications of environmental variability, heterogeneity and ecosystem complexity are discussed.

INTRODUCTION

The central question posed in this paper is almost philosophical. It is whether changes brought about by man's intervention in marine ecosystems are reversible. The question is central to the management of marine resources for on it depends the idea of the sustainability of harvesting. Yet, fundamentally, the question is abstract, for it involves the concept of determinism. Put simply, the composition of marine communities could be determined inexorably by the interactions of the component species or they could be produced by chance events in history, so that the observed composition is in fact but one of a set of equally possible structures (May, 1977).

If the question of reversibility is put in more practical terms, at the level of the individual species or stock, the potential productivity can clearly be eroded unequivocally by extinction. However, commercially exploited species have rarely been driven to extinction although the exceptions are instructive; usually such species have a low rate of increase, for example certain skates (Brander, 1982), or were particularly accessible to man such as certain marine mammals.

A more typical picture in recent times is of a pattern of overexploitation leading to a decline in the

9

stocks, followed by either a cessation or reduction in harvesting. The latter can occur either because the management authority demands it or economic factors make it necessary. At this level the question is whether, with reduced harvesting, the species can recover to original levels of abundance and productivity.

This question of reversibility may also be posed at the ecosystem level. A common pattern of exploitation involves the harvesting of a number of the component species of the ecosystem. The reasons may be practical, certain years involve a by-catch of different species; or economic, as clearly certain species are of commercial value while others are not. It is thus possible to query whether a whole community may be shifted towards a state in which valuable species form an insignificant part of the productivity of the system. The creation of such a 'marine desert' has usually been associated with forms of pollution. Nevertheless, the possibility exists that the harvesting of the community may lead to a reduction in most, if not all, valuable species. Indeed, such a situation appears to have happened in a number of tropical communities (Pauly and Murphy, 1982). An open question of key importance to the management of such fisheries is whether following reduction, or cessation, of harvesting the community would tend to return towards its original composition, abundance, and productivity.

Once the question is addressed in any detail, a whole number of qualifications become necessary. On an evolutionary time scale, it is clear that no community is unchanging. Environmental fluctuations produce corresponding fluctuations in component species of marine ecosystems; hence no community is, strictly speaking, unchanging. The patchiness of the marine environment implies that it is necessary to carefully define an appropriate spatial scale when discussing community composition. Finally, the structure of a community can be measured in a variety of ways; for example, as biomass, energy flow, or numbers of individual organisms.

The classical approach of ecological theory is to view ecosystems as represented mathematically by a set of equations describing the way in which the rates of change of component species are affected by their own abundance and that of the rest of the community (Lewontin, 1969). Such model systems may be represented deterministically by partial differential equations, which permit details of different life history types, ages, and sizes to be incorporated, as simple differential or difference equations, or as hybrids, incorporating explicit time delays. Environmental fluctuations may be incorporated by allowing parameters to vary stochastically and spatial patterns can be incorporated in a variety of ways, for example, by incorporating diffusion amongst patches, (Nisbett and Gurney, 1982). With this wealth of mathematical apparatus, it might be hoped that theory could act as a good guide to the likely response of different marine ecosystems to exploitation. This chapter examines this possibility, while reviewing briefly some of the empirical evidence that has accumulated about the way in which individual communities have responded to exploitation.

SINGLE SPECIES RECOVERY

Standard models of single species dynamics for fish typically only possess two equilibria: one at zero population size, the other at the carrying capacity. Additional equilibria are possible as a consequence of exploitation. The simple Schaefer model (Schaefer, 1957) and its variants (e.g., Fox, 1970), which specify the rate of change of stock biomass as a function of stock density all possess this characteristic. Similarly, models such as that of Beverton and Holt (1957) or dynamic pool models which include some stock and recruitment relationship and deal with the age composition of the stock also possess only a single, non-zero equilibrium (Pitcher and Hart, 1982). The predictions from such models concerning stock recovery are that provided the stock has not been harvested to extinction, a recovery towards the carrying capacity will occur as harvesting is reduced, or ceases.

There are some models which predict that below a certain theoretical density, extinction occurs. Such models are usually predicated on some model of mating behavior in which the birth rate falls below the death rate when mates become hard to find. Models of this type may well be applicable to marine mammals, but it is unlikely that, except at extremely low densities, they apply to fish. The absence of extinctions in commercially-exploited fish species, many of which have been dramatically reduced in abundance, points to the fact that such a situation may be rare. However, it is reasonable to examine the question whether, following a cessation in harvesting, recovery occurs towards the carrying capacity at an expected rate.

Unfortunately, this apparently simple question is difficult to answer in a quantitative way for most depleted fish stocks. The reason is that recruitment of young fish to the stock is highly stochastic (Hennemuth et al., 1980; Beddington and Cooke, 1983; Garrod and Colebrook, 1978). Hence, quantitative predictions require the calculation of the expectation of stock size at different times and the uncertainty associated with that expectation (Beddington, 1984).

This idea is elaborated in Figure 2.1 which illustrates conceptually the problems involved in assessing whether a particular stock has recovered at an expected rate, or not.

Empirical evidence, in part reviewed by Daan (1980), indicates that some species appear to have recovered at the expected rate; others have not. Clupeid species seem to be particularly variable in their behavior. Beddington et al. (1984) noted that even amongst a single herring species in the Northeast Pacific, certain stocks had stayed at low levels of abundance for long periods while others had apparently recovered from depletion at a rate close to the expectation (Kasahara, 1960). In the North Sea, the Downs Stock of herring has appeared to recover at a rate very similar to that which would be expected from the appropriate dynamic pool model (Beddington, 1980; Anon., 1983). Other stocks, for example the Georges Bank herring and Atlanto-Scandinavian herring have not. The reason for such different behavior is unknown, but it seems likely to depend on the behavior of the rest of the community.

It is clearly of relevance that for certain species there is a wealth of data to indicate quite dramatic changes in abundance in periods prior to significant levels of exploitation. The classic analysis by Soutar and Isaacs (1974) of the anaerobic sediments off California showed major changes in the abundance of pelagic fish in the last 2000 years. Similar work by De Vries and Pearcy (1983) on the Peruvian upwelling region shows a similar complicated pattern of shifting species abundance prior to exploitation. Such studies, for these systems at least, indicate the likelihood that alternative community compositions are possible. Indeed the California sardine is another stock whose recovery appears to have been delayed longer than would be predicted by simple single-species analysis (Murphy, 1977; MacCall, this volume).

An obvious implication of these results is that the simple single-species models are breaking down in certain circumstances, yet in others, appear to match well the behavior of exploited stocks. This is hardly helpful from a management perspective.

A possibility for extending simple single-species models without moving to a full representation of the dynamics of several species is by incorporating into the dynamics of the single-species, multispecies effects.

A key effect noted in many different ecological contexts is that of a predator whose functional response is sigmoid (Beddington, 1984). Imposing such a predation pattern on a simple production model produces the potential for two alternative stable states.

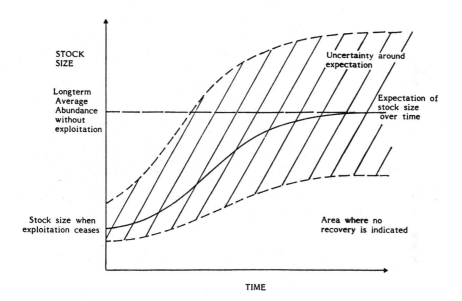

STOCK
SIZE

Longterm
Average
Abundance
without
exploitation

Uncertainty around
expectation

Expectation of
stock size
over time

Stock size when
exploitation ceases

Area where no
recovery is indicated

TIME

Figure 2.1. Schematic representation of stock recovery
when there is random variation in recruitment. The
uncertainty around the expectation is at a particular
probability level and the presence of the stock below
the lower bound of this region indicates that expected
recovery has failed to take place.

In such a model, high and low levels of predation produce only a single, near zero, equilibrium; intermediate levels produce the possibility of two. Steele and Henderson (1984) investigated such a model in which the level of predation varied stochastically. Unlike most other studies, Steele and Henderson's permitted the level of predation to be serially correlated. Their analysis indicated for different parameter combinations that the model was capable of mimicking, in qualitative terms at least, the sort of behavior associated with many stocks. As well as fluctuations around some equilibrium, dramatic changes in abundance could occur at quasi-regular intervals in the absence of exploitation. Similarly, a recovery from depletion could be delayed substantially, but then occur at a high rate. In general, they noted that an increase in fishing would enhance the tendency for the system to move between alternative states. Such analysis, even in the single-species case, has implications for management. Steele (1984) has even suggested that dramatic declines in certain exploited stocks may be caused by such phenomena rather than the more conventional explanation of overfishing.

Although incorporation of these complicating factors into single-species models increases the explanatory power, they give little guidance as to how to identify which stocks will behave in which way. Any hope of progress on such a problem will depend on the ability to successfully model other species in the system.

MULTISPECIES INTERACTIONS

In dynamic models of ecosystems, as the number of species increases, so does the propensity for there to be alternative stable states. The interaction of a prey-predator system considered above is just one mechanism which produces this potential. For example, if a further species is added as alternative prey to the predator, a whole set of new possibilities occurs, including the possibility that reduction of the predator will destabilize the interaction between the two competing prey (Hassell, 1978). Indeed, in many marine ecosystems where the same individual plays a different role as larvae, young fish, and mature fish, the possibilities for extremely complicated interactions abound. Hassell and Comins (1976) examined a relatively simple model where competition occurs at both juvenile and adult phases. They indicated that a variety of different stable states were possible, depending on the degree of competition at each life history stage. Although this model was aimed primarily as a description of the interaction of arthropods, there is a clear analogy with certain interactions of marine organisms and similar phenomena may be expected.

Given the complexity of most marine systems, it is quite clear that models of such systems are likely to possess an extremely complicated dynamic landscape with a substantial number of alternative states.

The recognition that such a situation is likely, means that it is extremely unlikely that there is going to be much potential for reversibility in complicated communities which have been harvested. Indeed, it may also be considered something of a paradox that some depleted stocks have recovered to something like those original abundances.

If some degree of stochasticity is introduced into complicated models, the time path of individual species abundance is likely to be quite intricate. Component species will move between alternative stable states as they are perturbed by environmental changes and when this factor is coupled with the effect of spatial heterogeneity on the marine environment, the potential to recognize alternate states in complex communities from observations on community abundance becomes minimal. The further implication, that only in relatively simple systems will different states be readily observed, seems to be borne out by such evidence that is available on community change (Beddington, 1984).

Although dynamic models of marine ecosystems appear to offer little in the way of guidance about the detailed behavior of communities, by contrast, empirical evidence appears to afford some insight. Brown et al. (1976) have examined changes in the stocks of the fish in the Northwest Atlantic as exploitation has proceeded. Their analysis indicates that although individual species are reduced, the overall biomass of the fish remains relatively constant. Explanations for such phenomena may be sought in common sense, rather than theory: prey species increase when predators are removed, one species benefits when a competitor is exploited. However, it is an open question to what extent such a generalization concerning the overall biomass applies in different dynamic systems.

Paine (1984) has recently reviewed a variety of approaches to modelling multispecies systems. He considers the classical multispecies approach reviewed in this chapter is a reasonable compromise between the single-species approach of traditional fisheries models and ecosystem models.

These latter are not considered in detail here, but Paine's dismissal of their utility seems valid. Paine argues that they are unable to predict the effect of exploitations on ecosystems for a variety of reasons. Of these, the most important in the context of this chapter is that such models aggregate species, often at the level of a guild. They are, thus, incapable of dealing with detailed changes within guilds or with the second order effects of the reduction of key species. Pimm (1982) has documented such second-order effects in a number of communities where a predator has been reduced. Of 19 studies reviewed by Pimm, only two showed little subsequent change in the species composition of the community, and several involved a species loss. Hence, such factors have to be recognized as important for management. The ecosystem models, in a sense, may be considered as providing little more than the assurance of common sense, backed by observation, that certain ecosystem properties exist under exploitation. They do not permit an evaluation at the species level, yet it is at this level that management problems are important. Similarly, on the central question of reversibility, such models offer no guidance.

CONCLUSION

An important implication of the inability of ecological theory to provide much in the way of insight into the response of complex communities to exploitation concerns the role of scientific data collection. If there is a common ground between observation and theory, it is at the multispecies level. These models which focus on specific ecological phenomena have less broad data demands than the ecosystem models, but are more demanding of detail on both temporal and spatial scales. If progress is to be made in understanding marine ecosystems, it is the view of this author that it will be achieved at this level. This implies that there is a need to monitor both on a long time scale and and on a large spatial scale the many components of the ecosystem. Such monitoring is rather unglamourous, but will be needed if alterations to the resource populations of large ecosystems are to be managed in the future, rather than endured.

ACKNOWLEDGMENT

This paper was prepared as part of the joint IIED/IUCN marine programme. Financial support of IUCN and the World Wildlife Fund is gratefully acknowledged.

REFERENCES

Anon. 1983. Report of the ICES Advisory Committee on fishery management. Copenhagen: ICES.
Beddington, J. R. 1980. Harvesting strategies for North Sea herring and the effect of random variation in recruitment. Report to European Economic Commission.

Beddington, J. R. 1984. The response of multispecies systems to perturbations. In Exploitation of marine communities. pp. 209-225. Ed. by. R. M. May. Springer-Verlag, Berlin.

Beddington, J. R. and Cooke, J. G. 1983. The potential yield of fish stocks. FAO Fish Tech. Paper 242.

Beddington, J. R., Arntz, W. E., Bailey, R. S., Brewer, G. D., Glantz, M. H., Laurec, A.J.Y., May, R. M., Nellon, W. P., Smetaeck, V. S., Thurow, F.R.M., Troadec, J.-P., and Walters, C. J. 1984. Management under uncertainty. Group Report. In Exploitation of marine communities. pp. 227-244. Ed. by R. M. May. Springer-Verlag, Berlin.

Beverton, R.J.H., and Holt, S. J. 1957. On the dynamics of exploited fish populations. Fish Invest. Ser. II, XIX, H.M.S.O., pp. 533.

Brander, K. 1982. Disappearance of common skate Raja batis from Irish Sea. Nature 290:48-49.

Brown, B. E., Brennan, J. A., Grosslein, M. D., Heyerdahl, E. G., and Hennemuth, R. C. 1976. The effect of fishing on the marine fish biomass in the northwest Atlantic from the Gulf of Maine to Cape Hatteras. ICNAF (Int. Comm. Northw. Atl. Fish.) Res. Bull. 12:49-68.

Daan, N. 1980. A review of replacement of depleted stocks by other species and the mechanisms underlying such replacement. Rapp. P.-v. Réun. Cons. int. Explor. Mer 177:405-421.

De Vries, J. G., and Pearcy, W. G. 1983. Fish debris in sediments of the upwelling area off central Peru: a late quarternary record. Deep Sea Res. 28:87-109.

Fox, W. W. 1970. An exponential surplus-yield model for optimising exploited fish populations. Trans. Am. Fish. Soc. 99:80-88.

Garrod, D. J., and Colebrook, J. M. 1978. Biological effects of variability in the north Atlantic Ocean. Rapp. P.-v. Réun. Cons. int. Explor. Mer 173:128-144.

Hassell, M. P. 1978. The dynamics of arthropod predator-prey systems. Monographs in Population Biology 13. Princeton Univ. Press, Princeton, N.J.

Hassell, M. P., and Comins, H. N. 1976. Discrete time models for two-species competition. Theor. Pop. Biol. 9:202-221.

Hennemuth, R. C., Palmer, J. E., and Brown, B. E. 1980. A statistical description of recruitment in eighteen selected fish stocks. J. Northw. Atl. Fish. Sci. 1:101-111.

Kasahara, H. 1960. Pacific herring. In H. R. MacMillon Lectures in Fisheries. Part 1. Fisheries Resources of the North Pacific Ocean.

pp. 49-63. Institute of Fisheries, U. British Columbia, Vancouver.

Lewontin, R. C. 1969. The meaning of stability. Brookhaven Symp. 22:13-24.

MacCall, A. D. This volume. Changes in the biomass of the California Current ecosystem.

May, R. M. 1977. Thresholds and breakpoints in ecosystems with a multiplicity of stable states. Nature 269:431-477.

Murphy, G. I. 1977. Clupeoids. In Fish population dynamics. Ed. by. J. A. Gulland. Wiley, N.Y.

Nisbett, R. M., and Gurney, W.S.C. 1982. Modelling fluctuating populations. Wiley, N.Y.

Paine, R. T. 1984. Some approaches to modelling multi-species systems. In Exploitation of marine communities. Ed. by R. M. May. Springer-Verlag, Berlin.

Pauly, D., and Murphy, G. I. 1982. Theory and management of tropical fisheries. ICLARM, Manila.

Pimm, S. L. 1982. Food webs. Chapman and Hall. London and N.Y.

Pitcher, A. J., and Hart, P.J.B. 1982. Fisheries biology. The AVI Publishing Co., London.

Schaefer, M. B. 1957. A study of the dynamics of the fishery for yellowfin tuna in the eastern tropical Pacific Ocean. Inter. Am. Trop. Tuna Comm. Bull. 2:245-285.

Soutar, A., and Isaacs, J. D. 1974. Abundance of pelagic fish during the 19th and 20th centuries as recorded in anaerobic sediments off California. Fish. Bull., U.S. 72:257-275.

Steele, J. H. 1984. Kinds of variability and uncertainty affecting fisheries. In Exploitation of marine communities. pp. 245-262. Ed. by R. M. May. Springer-Verlag, Berlin.

Steele, J. H., and Henderson, E. W. 1984. Modelling long term fluctuations in fish stocks. Science 224:985-987.

3. Long-Term Changes in the Baltic Ecosystem

ABSTRACT

The semi-enclosed Baltic Sea, 54° to 66°N, is divided into a series of basins, separated by shallows. It is connected to the Atlantic via a narrow and shallow transition. Exchange of deep and bottom water can only occur by water from the North Sea. The annual river runoff is about 2.2% of the volume, with large long-term fluctuations. A marked permanent salinity stratification results in a transition layer at 65 to 75 m. Large-scale meteorological conditions essentially determine the deep and bottom water exchange. There are considerable long-term fluctuations of salinity and temperature in the deep and bottom waters. The residence time of years to decades implies a gradual oxygen depletion. Anoxic conditions occur and the volume of anoxic water fluctuates greatly. During this century it has increased. This trend can be related to: climatic variations, an increased rate of oxygen consumption, an increased input of organic material from land based sources and from primary production, and a change in the ecosystem balance. The nutrient levels (phosphate, nitrogen) have increased over the last decades. A considerable input of various substances occurs through runoff and atmosphere. An increase is clear in some cases. During the mid-1940s, the cod stock increased. The herring and sprat stocks have also increased and the benthos has changed. Studies suggest that natural variability of water exchange and mixing are not main factors behind the changing conditions. Changes of oxygen consumption rates may be very important.

INTRODUCTION

Three factors make the Baltic Sea an exceptional area for investigations aimed at understanding natural

and man-made environmental changes: (1) the oceanographic characteristics; (2) the existence of long time
series of high quality observations of many physical,
chemical, and biological parameters over large parts of
the Baltic partly brought about by international cooperation so that the data are reasonably inter-
comparable; (3) the existence of information on inputs
of various substances, agreements towards their
control, and examples of effects of management and
control on localized scale.

Successful management requires an understanding of
the response of the system to various perturbations. A
central problem in our context is to distinguish between natural environmental variability and environmental variations caused by pollution. In this respect,
the Baltic Sea area has several important natural
characteristics: large positive water balance;
brackish water with large and stable horizontal and
vertical salinity gradients; strong physical constraints on water exchange and mixing, imposing large
residence times (several decades); special geochemical
systems through oscillations between oxic and anoxic
conditions in bottom and deep waters. Conflicting
human uses of the sea area include: waste disposal,
sea transportation, fishing, aquaculture, and
recreation. Large inputs of nutrients, organic material, and metals occur from land and atmosphere.

In the Scandinavian countries, populations and
industrial activities are concentrated along the coasts
of the Baltic Sea, with many small to medium size
harbors along the coast.

Along the eastern and southern borders of the sea
few, but large, population centers and harbors occur,
with generally less population concentration and industrial developments along the coasts (Bruneau, 1980).
However, large rivers enter the Baltic from a highly
agricultural drainage area along the southern coast.
During the last three decades marked agricultural
developments have occurred in the northern Baltic
countries. Today effective fertilization is applied so
that the production potential is well utilized.

The Baltic Sea ecosystem is influenced by a large
input of nutrients and other substances from a large
drainage area, as well as inputs from the atmosphere.
The whole system is coupled through interactions and
response times on various scales.

OCEANOGRAPHIC CHARACTERISTICS AND TRENDS

The semi-enclosed Baltic Sea (370,000 km^2; 21,000
km^3), 54°N to 66°N, an intracontinental Mediterranean
Sea, is one of the largest brackish water bodies in the
world. It is very shallow with a mean depth of 57 m

and about 17% of the area shallower than 10 m (e.g.,
Kullenberg, 1983). The Baltic Sea depression essentially constitutes a long fjord in the north-south
direction (~1,500 km) with an average width of
230 km. The topography divides the sea into a series
of relatively deep basins, with maximum depths in the
range of 105 m to 459 m. These are separated by sills
or shallows of depths in the range of 25 m to 140 m.
The sea is connected to the North Sea and the open
Atlantic via the Skagerrak and a narrow and shallow
transition area that includes the Kattegat, the Belt
Sea of sill depth 17 to 18 m, and the Øresund (Sound)
of sill depth 7 to 8 m. These topographic characteristics are of great importance for understanding the conditions in the Baltic Sea: a lateral exchange of deep
and bottom waters can only be accomplished by water
which has come from the North Sea, a distance of about
1,000 km, and that has passed over several shallow
sills.

The evolution of the Baltic Sea (see e.g., Winterhaller et al., 1981) has gone through several phases
since the last glaciation: the Baltic Ice Lake
(10,000-8000 B.C.), the Yoldia Sea (8000-7000 B.C.),
the Ancylus Lake (7000-5500 B.C.), and the Littorina
Sea (5500-1000 B.C.). At present the topography
changes through an annual uplift of 10 mm in the north
and subsidence of 1-2 mm in the transition area.

A second feature of the Baltic Sea is its marked
positive fresh water balance, with an annual river runoff of 440-480 km^3 or about 2.2% of the total volume
(Ehlin, 1981). The runoff usually has a maximum in May
and a minimum in January or February. It shows large
long-term variations, with amplitudes of 10-20% of the
long-term (100 yr) mean, primarily due to climatic
variability. The runoff generates an outgoing surface
layer flow of low salinity water, which sets up an
inward compensating flow at intermediate depths. A
marked permanent salinity stratification results in a
transition layer at 65 to 75 m.

A third feature of great importance is the influence of the meteorological conditions on the exchange
of the deep and bottom waters in the Baltic deep
basins. Favorable inflow situations occur with persisting westerly winds, high pressure over Jutland and
low pressure over Scandinavia. Often particularly
marked inflows occur at semi-periodic intervals of 3-5
yr, which can be related to specific, persisting
meteorological conditions over the Atlantic (Dickson,
1971). Highly saline water in the Baltic deep basins
will remain there until forced away by an inflow of
denser water. The bottom water salinity can only
decrease by vertical mixing which in the permanently
stratified Baltic Sea is a very slow process; a

decrease of about 1 ^{o}/oo takes about one year
(Kullenberg, 1982).
 Considerable long-term fluctuations of salinity
and temperature occur in the deep and bottom waters.
Trends have been established with increases of
temperature and salinity since the 1880s from 0.6 to
2.7°C and 0.8 to 1.7 ppt, respectively (Melvasalu et
al., 1981; Kullenberg, 1981; Matthäus, 1980, 1983a).
Fluctuations (periods of several years) of the salinity
are correlated with fluctuations in the river runoff,
at least in the 0-100 m part of the water column. The
trend of increasing salinity may be related to a long-
term decrease of the runoff of about 15% seen in observations from 1900-1960. However, over the period 1880
to 1980 no significant decrease of the runoff can be
seen. The increase of the salinity can be caused by
other factors, such as an increase of the salinity of
the inflowing water, a gradual increase of sill-depths
in the Danish Sounds due to the subsidence (Striggow,
1983), and a change in the water exchange through the
Sounds (Nehring, 1979). The latter may be due to a
change in the meteorological forcing of the water
exchange. The meteorological forcing shows large long-
term fluctuations (Kullenberg, 1977), but no significant trend can be seen over the period 1900-1960 from
direct wind observations. Pedersen (1982) found a
decreasing trend of the monthly peak-to-peak barometric
pressure at Bornholm over the period 1900-1970. The
observations clearly suggest the importance of the
river runoff, the wind, and the topographic constraints
for the oceanographic conditions in the Baltic Sea.
The wind provides the energy input for vertical mixing,
together with cooling in the fall-winter period, and
the energy for the lateral water exchange. The meteorological conditions over the Baltic Sea and the North
Sea regions are very important.
 The depth of the primary halocline also fluctuates
markedly in the range 20-30 m. However, the mean depth
has changed only slightly by 5-6 m from 77 m to 71 m
during the present century (Matthäus, 1980).
 The density of the water also displays long-term
trends of changes, with a general increase during the
present century (1900-1980) of about 0.5-1 σ_t unit
(Matthäus, 1983b). However, over shorter time periods
other trends can occur in different parts of the water
column. Matthäus (1983b) found a decrease of 0.4-
0.7 σ_t units at the 150-200 m level for the period
1952-1980, and an increase of 0.3-0.7 σ_t units in the
surface water during the same period. Large fluctuations of the stability over the primary halocline layer
occur (Fonselius, 1969; Kullenberg, 1977; Matthäus,
1983c). However, no significant long-term trend can be
seen (Kullenberg, 1981; Matthäus, 1983c). In the

deeper waters a decreasing trend is obvious during the present century (Matthäus, 1983c).

OXYGEN CONDITIONS

The long residence time, years to decades, of the deep and bottom waters in the Baltic Sea implies a gradual oxygen depletion due to oxidation of sinking organic matter and organic matter in the sediments. Anoxic conditions often occur in large parts of the bottom and deep waters (Fonselius, 1981). During this century there has been an increase of the volume of anoxic water, although interrupted by aperiodic decreases (Fonselius, 1969, 1981; Jansson, 1978). A temporary maximum occurred in the mid-1970s. The stagnant conditions usually last 3-5 yr, when the bottom waters are exchanged, a process which itself takes about 9 mo, operating successively in a cascade from the Bornholm Basin to the inner parts. Occasionally, after particularly strong or high saline inflows, the stagnant period, between inflows, can extend for 6 yr, which was the case for the last stagnation period. The length of these periods depends upon the density (salinity) of the water. Especially high salinities are found in the European shelf seas at intervals of 3-5 yr (Dickson, 1971).

A key question in relation to Baltic Sea management is why there is a trend of increasing oxygen depletion. The trend can be related to several factors:

(1) climatic variations, which influence the water balance and exchange, for instance giving rise to an increased residence time of the bottom waters. A decrease of the wind forcing, which influences both the mixing and the water exchange, is suggested from air pressure records since 1880; the decreasing trend is especially marked in the period 1900-1950 (Pedersen, 1982).

The depth of the halocline and the stability of the stratification across the halocline, both factors of importance for the mixing across the halocline, have not changed significantly. The mean circulation in the Baltic Sea is weak. It is conceivable that, given a decreasing energy input from the wind, the fluctuating circulation has decreased, which may imply increased residence times at intermediate levels (60-120 m).

(2) an increased rate of oxygen consumption, in bottom and deep water, which may be due to the increased temperature (Kullenberg, 1970), an increased amount of organic material in the water column (Shaffer, 1979), and a larger input of organic matter to the bottom due to an increased primary production (Jansson, 1978, 1984). This may lead to an accumulation of organic material in the sediments.

(3) an increased input of organic matter either from land-based sources and rivers or from an increase in primary production. Observations of primary production in different parts of the Baltic Sea since about 1960 suggest that an increase in primary production is occurring, amounting to about 50% over the last 20 yr (Kullenberg, 1983). The "average" primary production is about 100 g C m^{-2}yr^{-1}. An increase in the primary production by a factor 1.5-2 has also occurred in the parts of the Kattegat (Anon., 1984).

(4) a change in the ecosystem balance governing the uptake of nutrients and distribution of organic matter. Observations suggest that the occurrence of blue-green algae capable of uptake of atmospheric nitrogen has increased over the last 1-2 decades (Jansson, 1978, 1984). In recent years, several abnormal plankton blooms have occurred, but it is not possible as yet to conclude that the structure (length, timing) of the normal plankton blooms has changed, although there are indications of this. In the Baltic Sea there normally occurs a spring bloom in April or May, and a second bloom in July to August, whereas in the Gulf of Bothnia the spring and autumn blooms have merged (Jansson, 1978).

OTHER CHEMICAL CONDITIONS AND INPUTS

The level of phosphate in the surface layer winter water has increased by a factor of three over the last 2-3 decades (Nehring, 1974, 1981; Fonselius, 1981). The increase is also obvious in the deep water. Around 1969 the rate of increase of phosphate levels showed a marked increase, and since that year an increase of the nitrate concentrations has also occurred. The ratio N:P has decreased slightly (Nehring, 1984), which may imply that nitrogen is becoming increasingly important as the most limiting nutrient for primary production. An important question in respect to environmental management concerns the source of the nutrients. The N and P operate in different chemical cycles. In the case of nitrogen there exists the controlling factor of denitrification in oxygen depleted waters (Gundersen, 1981), which has been shown to be of great importance in the Baltic (Shaffer and Rönner, 1984). During anoxic conditions the phosphate which had been trapped in the sediments during oxic conditions is released to the water column. This is obvious from observations. When the deep and bottom waters are renewed, some phosphate is transferred towards the intermediate and surface layers. It is clear that considerable amounts of phosphate can reach the surface layer in this way.

A few large inflows from the Kattegat may generate the transfer required to explain the increase of phosphate levels in the surface winter water since 1968.

However, other very important sources of phosphate and nitrate are the rivers and the land runoff. Input studies suggest increases of these sources by several factors over the last decades (Jansson, 1984). These authors estimated annual river inputs of $50 \cdot 10^3$ and $640 \cdot 10^3$ tons of phosphorus and nitrogen, respectively, compared to $15-44 \cdot 10^3$ tons of phosphorus from the rivers estimated by ICES (1980) and Pawlak (1980). The recent estimates suggest an eight-fold increase of phosphorus and a four-fold increase of nitrogen inputs from land and atmosphere during this century (Jansson, 1984). The land-based inputs are very unevenly distributed and averages of concentration levels may not be appropriate. Clear trends of increases are seen in the central Baltic Basin and the Gdansk Basin. An increase has also occurred in the input of organic material. The atmospheric input is an important factor, amounting to about half the river input of nitrogen, for which positive trends have also been suggested (Anon., 1984; Jansson 1984). The nitrogen fixation by blue-green algae occurring during July-August also yielded a considerable input, estimated at $1-1.5 \cdot 10^5$ tons annually (Lindahl et al., 1977; Rinne et al., 1977; Gundersen, 1981).

Besides the input of nutrients and oxidizable organic material, large amounts of anthropogenic inputs of other materials occur. An increase in the sediments of 200% for Hg, and more than 100% for Cd, has been seen in many parts of the Baltic since about the time of the start of industrialization, 100-200 yr ago. Increases have also been established for Pb, Zn, and Cu (Brügmann, 1981). The geochemistry of these metals imply that large amounts of the land-based inputs are trapped in the coastal zone sediments, whereas the often significant atmospheric inputs are distributed all over the basin.

Substances like DDT, PCB, and toxaphene are also present in large amounts in different parts of the Baltic system, including the biota (ICES 1977; Dybern and Fonselius, 1981). Reproduction failures in Baltic marine mammals have been ascribed to pollution by DDT and PCB. The substances DDT, PCB, and Hg provide good examples of the possibilities of control and management of inputs. Since the use of these materials was limited or prohibited by law in several countries bordering the Baltic, the concentrations in biota have decreased in many areas (Dybern and Fonselius, 1981; Ehrhardt, 1981; Helle and Stenman, 1984); despite this the levels in seals are still quite high.

ASPECTS OF THE ECOSYSTEM

The organisms in the Baltic Sea are adjusted to low salinities. There are fewer species, they are less specialized and smaller than in fully marine areas (Jansson, 1978; Kullenberg, 1983). Temperature governs the seasonal and vertical variation of the zooplankton fauna, and salinity affects the species composition. The ecosystem is complex and is subject to large natural and man-made variations, but it is very difficult to ascertain the impacts of the latter on the system. It appears that a change has occurred from oligotrophic to eutrophic conditions in coastal zones and possibly parts of the open sea (Dybern and Fonselius, 1981; Jansson, 1984). The benthos has shown considerable changes over the last 50 yr. There are indications that oligotrophic fish such as pike and perch are being replaced by the eutrophic bream and roach (Jansson, 1978). In some coastal areas the Fucus has decreased or disappeared, whereas green and brown algae (Cladophora and Enteromorpha) have increased.

The increased primary production has led to an increasing sedimentation and accumulation, slowly changing hard bottoms to soft bottoms (Jansson, 1984). Large-scale, long-term changes in the benthic soft bottom macrofauna have been observed in deep parts of the Baltic (Bornholm-Gdansk-Central basins and Gulf of Finland), amounting to a decline or disappearance of macrofauna or a change of the fauna composition (Andersin et al., 1978). The changes started in the early 1950s in the southern parts, and in the late 1950s in the northern parts (Andersin et al., 1978). The changes were partly triggered by the very large salt water inflow in 1951-1952, described by Wyrtki (1954).

Cederwall and Elmgren (1980) presented results from benthic macrofauna biomass investigations repeated in the late 1970s at the same stations as had been occupied in the 1920s. Above the halocline layer a large increase in biomass was established whereas in areas below the halocline depth (~70 m) a very strong decrease in biomass was established. The average annual increase above was 2-4% and Cederwall and Elmgren (1980) judged this to be due to eutrophication leading to a higher food supply to the benthos than before. The long-term changes in natural oceanographic conditions (S, T, stability, etc.) have not been large enough to explain the drastic changes in various parts of the ecosystem.

The fish catch has increased steadily since the 1930s, herring, sprat and cod being the most important species. During the mid-1940s the cod stocks increased. Since 1977 this species has shown another

increase and the cod stocks are now stronger than ever
(Otterlind, 1983; Jansson, 1984). The total fish catch
has doubled during the last 20 yr, and is now about 0.9
million tons annually. It is possible that an increase
in herring and sprat stocks also started before the
increase of anthropogenic inputs during the 1950s and
1960s, but it seems likely that the fish stocks have
benefited from the increased input of nutrients
(Otterlind, 1983).

During recent years frequent observations have
been made of various types of fish diseases in many
parts of the Baltic Sea. It is not possible to
demonstrate to what extent this is due to contamination, except in some localized cases where industrial
inputs probably have had an influence.

MANAGEMENT AND CONTROL

Despite the existence of long time series of
observations and some insight as to the forcing functions governing the variability, we still do not
understand the dynamics of the system well enough to
clearly explain the long-term changes or to be able to
make reliable predictions of effects of changes of
inputs of substances, changes of river runoff or
changes of the topographic conditions. In order to
achieve this we need a more thorough understanding of
the processes and cycles. As far as nutrients are
concerned, we need to know more about the nitrogen
cycle and the relative importance of phosphate and
nitrate under different conditions. In relation to
inputs from land generally, further information is
necessary on the role of the coastal zone as a filter
or trap. This could be achieved partly through observations in sections from the river, across the estuary
and the coastal zone to the open sea. In relation to
ecosystem effects of contamination, or pollution, we
need to establish dose-response and cause-effect relationships, which could be done through mesocosm and
laboratory experiments, using realistic concentration
levels. Biological effects studies need to be
encouraged, and there is also a need for exchange of
information on observed effects in the field and
coordination of such studies.

We also need to know the response time of the
system to rather marked natural events, like the so-
called mid-1970s salinity temperature anomaly which can
be followed over most parts of the North Atlantic. Can
the long duration of the last stagnation period be
related to this anomaly? Dynamical models using
realistic forcing functions need to be developed which
can be used to investigate the response of the system
to various perturbations.

Quite clearly there are cases where information about deteriorating conditions and effects of pollution can be, or have been, established and also may be related to given sources of contamination. Such cases call for action limiting the inputs, control and monitoring, so as to establish the results of the management. Often these cases are localized. Exceptions are the cases of DDT, PCB, and Hg, where the source functions were not very well defined. Studies of the localized cases should also be encouraged with a view to show the effect of management, and to more clearly establish the limits of adjustment. Do we need a complete ban of a substance? This usually leads to the introduction of some substitute, which also may turn out to be very undesirable from an environmental point of view.

The outlook for maintaining cognizance of the pollution problem in the Baltic ecosystem and taking steps to reduce adverse ecological effects is good. Several international scientific organizations, and one international commission, with membership from Baltic coastal countries have ongoing programs to monitor changes of key ecosystem components. A Baltic monitoring program is reporting results to the Helsinki Commission, which is providing management advice to member nations. Requests of the Helsinki Commission have been directed to the International Council for the Exploration of the Sea for specific assessments of contaminant effects. The results of multinational and multidisciplinary studies of pollution effects on the Baltic ecosystem were reported recently at the 14th Conference of Baltic Oceanographers held in Gdynia, Poland, in August 1984. An active program of measuring pollution effects and natural environmental changes on the productivity of Baltic living marine resources is conducted by ICES. Baltic marine biologists conducted a symposium on "The Ecology of Coastal Waters in the Baltic Sea" in Turku, Finland, during summer 1985. The growing concern over the apparent ecological changes in the Baltic resulted in a significant augmentation of studies of the Baltic during the second International Baltic Year (IBY-2) during the 12-month period, July 1985 through July 1986. Letters addressed to countries around the Baltic were sent by ICES to encourage their support in the research projects associated with IBY-2. Much of the recent and planned work for the mid-1980s is summarized in the Report of the ICES/SCOR Working Group on the Study of Pollution of the Baltic [(SCOR WG 42) Helsinki April 1985] and in the annual meeting reports of ICES available from the Secretariat in Copenhagen.

REFERENCES

Andersin, A.-B., Lassig, J., Parkkonen, L., and Sand-
ler, H. 1978. The decline of macrofauna in the
deeper parts of the Baltic proper and the Gulf of
Finland, Kiel. Meeresforsch. Sonderheft 4:23-52.
Anonymous. 1984. Iltsvind og Fiskedød i 1981. Omfang
og Årsager. In Danish; Miljøstyrelsen, Strandgade
29, 1401 Copenhagen K. 247 pp.
Brügmann, L. 1981. Heavy metals in the Baltic Sea.
In The State of the Baltic, Ed. by G. Kullenberg,
Mar. Pollut. Bull. 12 (6):214-218.
Bruneau, L. 1980. Pollution from industries in the
drainage area of the Baltic. Ambio, Special issue
on the Baltic p. 145-152.
Cederwall, H., and Elmgren, R. 1980. Biomass increase
of benthic macrofauna demonstrates eutrophication
of the Baltic Sea. Ophelia, Suppl. 1:287-304.
Dickson, R. R. 1971. A recurrent and persistent
pressure-anomaly pattern as the principal cause of
intermediate-scale hydrographic variation in the
European shelf seas. Dtsch. Hydrogr. Z. 24(3):97-
119.
Dybern, B. I., and Fonselius, S. H. 1981. Pollution.
In The Baltic Sea, pp. 351-382. Ed. by A. Voipio.
Elsevier Oceanography Series 30, Amsterdam.
418 pp.
Ehlin, U. 1981. Hydrology of the Baltic Sea. In The
Baltic Sea, pp. 123-134. Ed. by A. Voipio.
Elsevier Oceanography Series 30, Amsterdam.
418 pp.
Ehrhardt, M. 1981. Organic substances in the Baltic
Sea. In The State of the Baltic, Ed. by G.
Kullenberg, Mar. Pollut. Bull. 12(6):210-213.
Fonselius, S. H. 1969. Hydrography of the Baltic deep
basins III. Fish. Board Swed. Ser. Hydrogr. 23:97.
Fonselius, S. H. 1978. On nutrients and their role as
production limiting factors in the Baltic. Acta
Hydrochem. Hydrobiol. 6:329-339.
Fonselius, S. H. 1981. Oxygen and Hydrogen Sulphide
Conditions in the Baltic Sea. In The State of the
Baltic, Ed. by G. Kullenberg, Mar. Pollut. Bull.
12(6):187-194.
Gundersen, K. 1981. The distribution and biological
transformations of nitrogen in the Baltic Sea. In
The State of the Baltic, Ed. by G. Kullenberg,
Mar. Pollut. Bull. 12(6):199-205
Helle, E., and Stenman, O. 1984. Recent trends in
levels of PCBs and DDT compounds in seals from the
Finnish waters of the Baltic Sea. ICES C.M.
1984/E:3, 8 pp. (mimeo).

30

ICES. 1977. Studies of the pollution of the Baltic Sea. ICES/SCOR Working Group on the Study of Pollution of the Baltic Cooperative Research Report 63, ICES, Copenhagen, 97 pp.

ICES. 1980. Assessment of the marine environment of the Baltic Sea Cooperative Research Report XX. ICES, Copenhagen XX pp.

Jansson, B. O. 1978. The Baltic--A systems analysis of a semi-enclosed sea. In Advances in Oceanography, pp. 131-184. Ed. by H. Charnock and G. Deacon, Plenum Press, Oxford.

Jansson, B. O. 1984. The Baltic Sea and the nutrients. In The Baltic Sea--an environment worth protecting. Symposium in Karlskrona, Sweden. June 1984. (In press)

Kullenberg, G. 1970. On the oxygen deficit in the Baltic deep water. Tellus 22:357.

Kullenberg, G. 1977. Observations of the mixing in the Baltic thermo- and halocline layers. Tellus 29(6):572-587.

Kullenberg, G. 1981. Physical Oceanography. In The Baltic Sea, pp. 351-382. Ed. by A. Voipio. Elsevier Oceanography Series 30, Amsterdam, 418 pp.

Kullenberg, G. 1982. Mixing in the Baltic Sea and implications for the environmental conditions. In Hydrodynamics of semi-enclosed seas, pp. 399-418. Ed. by J.C.J. Nihoul. Elsevier Oceanography Series 34, Amsterdam, 555 pp.

Kullenberg, G. 1983. The Baltic Sea. In Estuaries and Enclosed Seas, Ed. by B. H. Ketchum, vol. 26. Ecosystems of the World, Elsevier, Amsterdam, 500 pp.

Lindahl, G., K. Wallström, and Brattberg, G. 1977. On nitrogen fixation in a coastal area of the northern Baltic. BMB 5th Symposium, Kiel, 1977. 6 pp. (unpublished manuscript).

Matthäus, W. 1980. Zur Variabilität der primären halinen Sprungschicht in der Gotlandsee. Beitr. Meereskd., Berlin 44/45:27-42.

Matthäus, W. 1983a. Aktuelle Trends in der Entwickling des Temperatur-, Salzgehalts-, and Sauerstoffregimes im Tiefenwasser der Ostsee. Beitr. Meereskd., Berlin 49:47-64.

Matthäus, W. 1983b. Langzeittrends der Dichte im Gotlandbecken. Beitr. Meereskd., Berlin 48:47-56.

Matthäus, W. 1983c. Zur Variationen der vertikalen Stabilität der thermohalinen Schichtung im Gotlandtief. Beitr. Meerskd., Berlin 48:57-71.

Melvasalu, T., Pawlak, J., Grasshoff, K., Thorell, L., and Tsiban, A. (Editors). 1981. Assessment of the effects of pollution on the natural resources of the Baltic Sea, 1980. Baltic Sea Environment Proceedings Vol. 5A and 5B, Helsinki.

Nehring, D. 1974. Untersuchungen zum Problem der Denitrifikation und Stickstoffenbindung im Tiefenwasser der Ostsee. Beitr. Meereskd., Berlin, 33:135-139.

Nehring, D. 1979. Relationships between salinity and increasing nutrient concentration in the mixed winter surface layer of the Baltic from 1969 to 1978. ICES C.M. 1979/C:24 (mimeo) 8 pp. (See also Annales Biologique 1978).

Nehring, D. 1981. Phosphorus in the Baltic Sea. In The State of the Baltic, Ed. by G. Kullenberg, Mar. Pollut. Bull. 12(6):194-198.

Nehring, D. 1984. Essential nutrients. First draft for Baltic Sea Assessment 1984. Helsinki Commission.

Otterlind, G. 1983. Östersjön och övergödningen. Yrkesfiskaren no. 18-20, in Swedish. Marine Research Laboratory, 45300 Lysekil, Sweden.

Pawlak, J. 1980. Land-based inputs of some major pollutants to the Baltic Sea. Ambio special issue on the Baltic, pp. 163-167.

Pedersen, F. B. 1982. The sensitivity of the Baltic Sea to natural and man-made impact. In Hydrodynamics of semi-enclosed seas, pp. 385-399. Ed. by J.C.J. Nihoul. Elsevier Oceanography Series 24, Amsterdam, 555 pp.

Rinne, I., Melvasalu, T., Niemi, Å, and Niemisto, I. 1977. Nitrogen fixation by blue-green algae in the Baltic Sea. BMB 5th Symposium, Kiel, 7 pp. (Unpublished manuscript)

Shaffer, G. 1979. On the phosphorus and oxygen dynamics of the Baltic Sea. Contributions from the Askö Laboratory, 26:90

Shaffer, G., and Rönner, U. 1984. Denitrification in the Baltic proper deep water. Deep-Sea Res. 31:197-220.

Striggow, K. 1983. Die relative Landsenkung im Bereich des Sunds und der Beltsee--eine weitere Ursache der rezenten Salzgehaltszunahme der Ostsee. Gerlands Beitr. Geophys. 92 (2-4):228-240.

Winterhaller, B., Flodén, T., Ignatius, H., Axberg, S., and Niemisto, L. 1981. Geology of the Baltic Sea. In The Baltic Sea, pp. 1-122. Ed. by A. Voipio. Elsevier Oceanography Series 30, Amsterdam, 418 pp.

Wyrtki, K. 1954. Der grosse Salzeinbruch in die Ostsee im November und Dezember 1951. Kieler Meeresforsch. 10:19-25.

Alec D. MacCall

4. Changes in the Biomass of the California Current Ecosystem

ABSTRACT

California Current pelagic fishes have been monitored for 30 to 50 years, and a paleosedimentary record extends back 200 to 2000 years. Large natural fluctuations in abundance occur at all time scales. Overharvest of sardines removed a major component of the ecosystem; the extent to which it was replaced by other species (e.g., anchovy, rockfish) is not clear. Large predatory fishes have declined in abundance due to exploitation. Pinnipeds were depleted in the last century, but now are abundant and are increasing rapidly. Seabird reproductive success is closely related to availability of small forage fish such as anchovy. Despite a wealth of scientific information, species interactions are poorly understood, and are difficult to separate from independent differential responses to varying environmental conditions. The biological basis of fishery management is likely to remain single-species models in the foreseeable future. Ecosystem management requires coordinated consideration of both fished and non-fished species, but faces conflicting jurisdictions and other institutional difficulties.

INTRODUCTION

The California Current sweeps southward along the west coast of North America, with its main influence extending from about the Columbia River to central Baja California (Figure 4.1). A major eddy system occurs in the Southern California Bight (SCB) south of Point Conception, and a second eddy system occurs in Sebastian Vizcaino Bay, Baja California. Wind-driven coastal upwelling occurs along the more exposed sections of coastline, particularly at headlands, cooling the inshore area and adding nutrients to the already rich

33

34

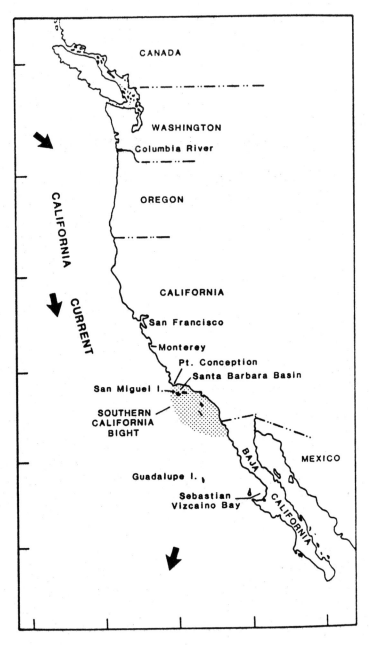

Figure 4.1. Map of the California Current region, with some place names which are referred to in the text.

waters being advected southward from the northeast Pacific. The SCB is a special habitat within this system, being warmer and influenced by numerous islands and banks. Most of the remainder of the California Current has little interaction with the seabed due to a relatively narrow continental shelf. Some useful reviews of California Current oceanography are Reid et al. (1958), Hickey (1979), and Parrish et al. (1983).

The main body of the California Current lies offshore and contains an abundance of lower trophic level organisms; this area has been studied extensively by biological oceanographers who have benefited from regular surveys made by the California Cooperative Oceanic Fisheries Investigations (CalCOFI) since 1950. From the viewpoint of this volume, more emphasis must be placed on the inshore regions, especially the SCB, where higher trophic level organisms such as fish, marine mammals, and seabirds abound. It is these organisms that usually are of most concern to management, as they play a visible role in human culture and economy.

PREINDUSTRIAL CONDITIONS

The natural state of the California Current ecosystem was almost certainly one of substantial variability at all time scales. Anaerobic sediments in the Santa Barbara Basin of the SCB contain a rich record of recent fossil remains, particularly fish scales from the past few centuries. The annual layering of these sediments allows precise dating, and enabled Soutar and Isaacs (1969, 1974) to compile time series of scale deposition rates which may be interpreted as rough indexes of abundance (Figure 4.2). While the process of scale deposition and relation to the source population is poorly defined, scale deposition rates tend to agree with well-documented changes in commercial fish abundance since 1930, providing some credibility for their interpretation (Soutar and Isaacs, 1974; Smith, 1978; Lasker and MacCall, 1983).

Another demonstration of long-term variability is given by Hubbs (1948), who interprets lists of fish species collected by the Pacific Railroad Survey in 1853-1857 and by Jordan and Gilbert in 1880. Many tropical species are shown which are exceedingly rare or unknown in modern experience; other species normally restricted to the SCB and southward are reported as being abundant in Monterey Bay, central California. Douglas (1980) has used tree-ring data to reconstruct past sea surface temperatures off California, and the period 1841-1859 appears anomalously warm. Northward faunal shifts are known to be associated with oceanic warming off Califonia (Hubbs, 1948; Radovich 1961), but

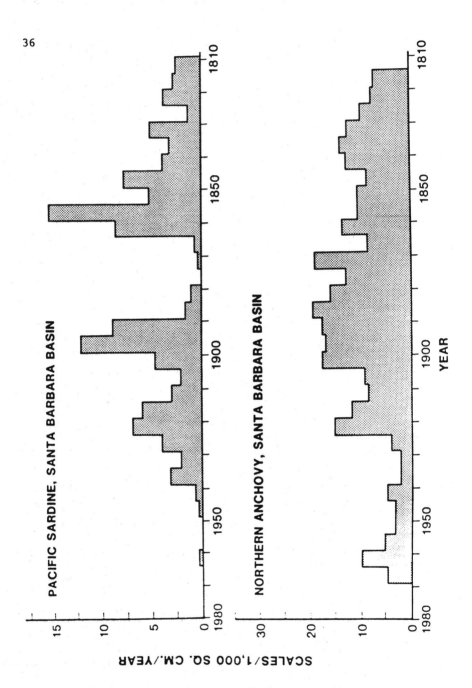

Figure 4.2. Recent scale deposition rates for Pacific sardine and northern anchovy (from Soutar and Isaacs, 1974).

nothing of the magnitude and duration experienced in the mid-1800s has been seen since then.

Marine mammal abundances were severely impacted by the influx of Europeans in the 1800s. The major fur bearers--northern fur seal (Callorhinus ursinus), Guadalupe fur seal (Arctocephalus townsendii), and California sea otter (Enhydra lutris)--were quickly reduced to near-extinction, as were the northern elephant seal (Mirounga angustirostris) and the California gray whale (Eschrichtius robustus) which were rendered for their oil. Other pinnipeds less sought for commercial purposes, such as the California sea lion (Zalophus californianus), suffered from prolonged hunting before concern turned to predator control for the marine fisheries.

Seabirds also were impacted by the influx of Europeans, particularly following the gold rush of 1849. Seabird eggs were harvested for food, and disturbance of the breeding colonies caused further reproductive difficulty. In 1854 common murres (Uria aalge) numbered about 400,000 on the Farallon Islands near San Francisco, judging by the magnitude of the annual egg harvest (Ainley and Lewis, 1974). As recently as 1959 these birds numbered only about 6,000, but their population has since recovered to 60,000 as a result of protection from disturbance (Sowls et al., 1980). Changes of seabird abundance in other locations are poorly documented, but most accessible colonies near human population centers were probably adversely impacted.

Except for the few fishes represented in Soutar and Isaacs' sedimentary record, it is particularly difficult to reconstruct the prehistoric state of the California Current ecosystem. Abundances of animal remains in prehistoric middens suggest that the Guadalupe fur seal may have been the dominant pinniped in southern California (Walker and Craig, 1979). Ironically this species does not presently reproduce in southern California although it has recovered somewhat on Guadalupe Island where it was "rediscovered" only 30 years ago (Antonelis and Fiscus, 1980). The northern fur seal has a southern rookery on San Miguel Island where waters are cool due to strong upwelling off Pt. Conception. Their prehistoric abundance at this location may have been high, and they probably were quite important in the northern part of the California Current, with large numbers of subadult females migrating southward from the Pribilof Islands. A major peak in Soutar and Isaacs' (1974) scale deposition rate for Pacific hake (Merluccius productus) coincides with peak harvests of fur seals at the Pribilof Islands in the late 1800s (Anon., 1977).

38

Aboriginal impacts on natural resources are usually thought of as being in approximate equilibrium. McEvoy (1979) concludes that aboriginal salmon harvests in northern California approached maximum sustainable yield (regulated by a variety of cultural mechanisms) and greatly exceeded harvests sustained by later industrial fisheries. Impacts on most other marine species were relatively small. An apparent exception is Chendytes lawi, a flightless marine duck with a well documented evolutionary lineage over the last 2 million years in California (Warter, 1978). The most recent remains from Indian middens have been dated at less than 3,780 years B.P. (Morejohn, 1976). As Morejohn observes:

> The high frequency of occurrence of bones of this species at one Indian midden clearly implicates early California aboriginal man as playing an important role in its extinction.

RESOURCE VARIABILITY IN THE 20th CENTURY

Commercial fisheries in California have been studied and monitored closely since the early part of this century, but they have been managed very little. Recreational fisheries, which have a major impact on many California fish stocks, have been monitored since the late 1940s, although substantial fishing pressure has existed for much longer. Mexican fisheries probably had a small impact on California Current fish stocks prior to the 1960s, and some segments of that industry have grown substantially in the last decade. Marine mammals and seabirds were not extensively monitored or studied until the 1970s when public awareness increased and several environmentally oriented acts were promulgated by the U.S. and California legislatures. This section discusses the variability of these California Current resources, and the subsequent section discusses their management.

The collapse of California's Pacific sardine (Sardinops sagax caerulea) fishery in the late 1940s set a well-documented pattern later followed by many of the world's coastal pelagic fisheries. However, the appearance of collapse may be an artifact of scale: A logarithmic plot of historical sardine abundances (Figure 4.3) indicates a prolonged exponential decline with a remarkably constant average rate of decay, about -0.2/yr. MacCall (1983) interprets this as a constant average rate of overexploitation, with harvests exceeding replacement levels by about 18% over the entire history of the fishery. Also, the variability about the trend in Figure 4.3 is of roughly constant magnitude indicating constant proportional variability

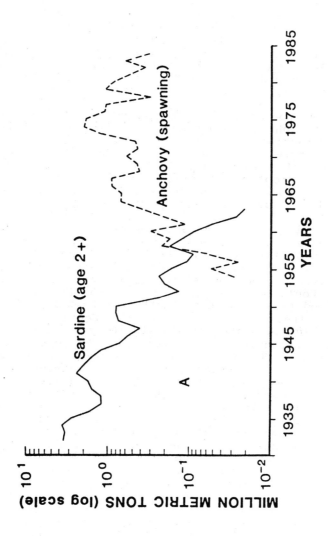

Figure 4.3. Time series of sardine (age 2+) and anchovy spawning biomass (log scale) off California and northern Baja California. "A" denotes approximate anchovy spawning biomass in 1940-1941 (interpreted from Smith, 1972). Sardine biomasses are from Murphy (1966) and MacCall (1979); anchovy biomass are from MacCall and Methot (1983).

from environmental fluctuations, and suggesting that difficulties experienced near the end of the fishery were not unusually severe. Removal of a major component such as the Pacific sardine (unexploited biomass averaging about 3 million tons) from the California Current food web would presumably have measurable effects on other components of the ecosystem. It is surprisingly difficult to identify those putative effects.

The northern anchovy (Engraulis mordax) is commonly thought to be a competitor with the sardine. The circumstantial evidence of similar food habits and a large increase in anchovy abundance following the decline of the sardine (Figure 4.3) has often been cited as proof of competitive replacement (e.g., Odum, 1971). However, the increase in anchovy abundance lagged the sardine decline by nearly a decade, and Smith's (1972) evidence of a relatively large anchovy biomass in 1940, when sardines were abundant, is often overlooked. A direct interaction between the two species has been demonstrated by Santander et al. (1983) in Peru, where anchovies and sardines are mutual predators on their eggs and larvae; a similar relationship should occur in California. Possible competition for food is suggested by Lasker and MacCall (1983), who showed that anchovy scales from Soutar and Isaacs' anaerobic sediments are significantly smaller (implying smaller anchovies) when sardine scale deposition rate is high. However, recent observations weaken the hypothesis of strong interaction between the two species: Mean size of anchovies has declined drastically since the mid-1970s due to causes that cannot be attributed entirely to the increased fishing pressure. At the same time, Pacific sardines are showing signs of increase, but are still very scarce. If sardines were more abundant, the increase in sardine abundance would be thought to be the cause of the decline in anchovy mean size, but because sardines are presently an insignificant component of the trophic system, these phenomena must be independent results of another, presumably environmental influence. As Daan (1980) concluded in a general review of replacement phenomena, the case is very unclear for anchovy having competitively replaced the sardine in California.

Due to a string of very strong year classes, abundance of Pacific mackerel (Scomber japonicus), elsewhere known as chub mackerel, in the 1930s may have greatly exceeded average virgin stock levels (Figure 4.4). Over the following years spawning success (recruits per spawner) varied with a distinct 6- to 7-yr cycle which ended with severe depletion of the resource about 1965. A moratorium on fishing during the 1970s was successful in rehabilitating the resource, which

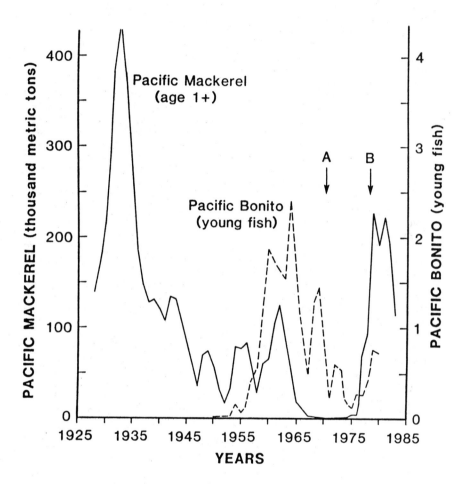

Figure 4.4. Time series of Pacific mackerel total biomass (age 1+) and Pacific bonito recreational catch per effort (fish per angler-trip). "A" denotes beginning of Pacific mackerel fishing moratorium, and "B" denotes re-opening of the fishery. Pacific mackerel biomasses to 1969 are from Parrish and MacCall (1978); bonito catch per effort is from Collins et al. (1980).

rebounded by producing the largest year class (1978) in the history of the fishery (Klingbeil, 1983). Surprisingly, the cyclic spawning success seen consistently in the 30 yr prior to collapse has not been evident in the erratic pattern of the recovered resource.

Pacific bonito (Sarda chiliensis) have long been seasonal visitors to California waters, with large northwardly migrating adults appearing in the SCB in the late summer and fall. However, an apparently self-sustaining population seems to have established residence in California only since the late 1950s, perhaps as a result of the oceanic warming of 1958-1959 (see Collins et al., 1980, for a complete review). Abundance has declined since the early 1960s (Figure 4.4). Some of that decline must be attributed to the commercial fishery which began harvesting bonito in 1966, but sustainability cannot be expected from a resource which has only recently come into existence. The bonito stock in the SCB seems to be a marginal population that hovers near the edge of viability. One hypothesis for its existence is that warm-water discharges from power generating stations have provided overwintering refuges for fish that would normally migrate southward (Collins and MacCall, 1977).

No substantial fishery has developed for jack mackerel (Trachurus symmetricus) in California or Mexico. A small fishery for young fish has existed since 1950 but the industry has had difficulty marketing the product. Much of the biomass consists of large adults inhabiting oceanic waters from the tip of Baja California to the Aleutian Islands, and as much as 1500 km offshore. These fish may exceed 30 yr of age, and stock size may be around two million tons (MacCall and Stauffer, 1983). Nearly all the spawning occurs in offshore oceanic waters, but subsequent young fish appear mainly in the SCB which serves as a nursery ground for a few years. Recruitment strength is highly variable (Figure 4.5), with seemingly periodic strong year classes tending to be isolated by several weak year classes. Age compositions for recent catches are unavailable, but the 1976 year class seems to be prominent, maintaining the pattern.

There are about 60 known species of rockfishes (Sebastes spp.) in the California Current, most of which are demersal and reside north of the U.S.-Mexican boundary (Miller and Lea, 1972). Some of the more pelagic rockfishes such as shortbelly (S. jordani), bocaccio (S. paucispinis), and chilipepper (S. goodei) potentially interact with shallow schooling pelagic fishes such as anchovy and sardine. Abundance of rockfish larvae has been monitored by the CalCOFI ichthyoplankton surveys since the early 1950s, and a few species such as shortbelly and bocaccio are

43

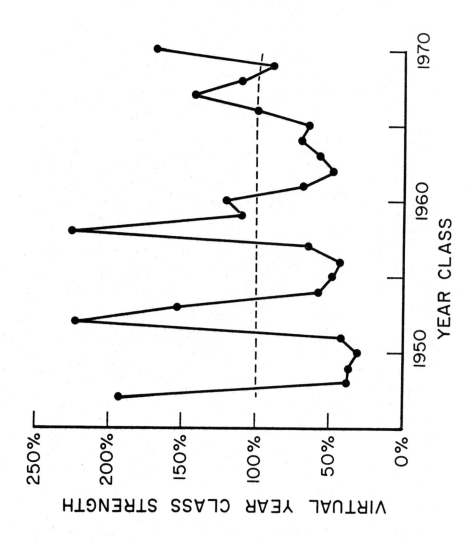

Figure 4.5. Relative recruitment strengths of jack mackerel year classes in southern California (From MacCall and Stauffer, 1983).

identifiable as small larvae. Abundance of adult
shortbelly rockfish appears to be several hundred
thousand tons, based on acoustic-trawl surveys
(inferred from Lenarz, 1980), and, as this species
contributes about one-tenth of all Sebastes larvae in
the CalCOFI surveys, total rockfish abundance must
amount to several million tons. Relatively few species
of rockfish are sufficiently abundant, vulnerable, and
desirable, to be the object of major commercial or
recreational fisheries.

Abundance of rockfish larvae has varied substantially over the last 30 yr. In southern California,
where sampling has been the most consistent, the data
show three distinct periods (Figure 4.6). The early
and recent periods appear fairly stable, with a suggestion of higher abundance recently. The anomalous
middle period is initiated by the oceanic warming of
1958-1959 when larval abundances dropped severely.
Unlike most other fish species, rockfish larval abundances did not recover immediately following the return
to normal oceanic conditions in the early 1960s, but
took about ten years to recover to previous levels. It
is unclear to what extent the depressed larval abundances were due to decreased adult abundance or to
decreased fecundity of those adults. There is no trend
in shortbelly rockfish larvae (Figure 4.6) despite
large changes in abundance of potentially competing
anchovy and sardine. Abundance of bocaccio larvae
(Figure 4.6) seems to parallel changes in abundance of
anchovy, but the reason is not known; however, adult
bocaccio are large fish, and are probably predators of
anchovy.

Most of the larger predatory fishes are migratory.
The temperate tunas, bluefin (Thunnus thynnus) and
albacore (T. alalunga) visit the California Current as
1- to 3-yr-old fish in late summer during their trans-
pacific migrations. Catch rates of both species have
declined from peak levels of the 1960s (Hanan, 1983;
Laurs, 1983). Many predatory fishes migrate along the
coast, wintering in Baja California waters, and moving
northward into California waters in the summer. Most
of these species are now at low abundance relative to
those of the early 1900s. Intense commercial and
recreational fisheries on white seabass (Atractoscion
nobilis) and California barracuda (Sphyraena argentea)
have resulted in very low abundance of these species.
A contraction of range is evident for white seabass,
which formerly supported a commercial fishery in San
Francisco, but now occurs only south of Pt. Conception
(Methot, 1983; also see reviews by Collins, 1981; and
Vojkovich and Reed, 1983, for white seabass; and
Schultze, 1983, for barracuda). Yellowtail (Seriola

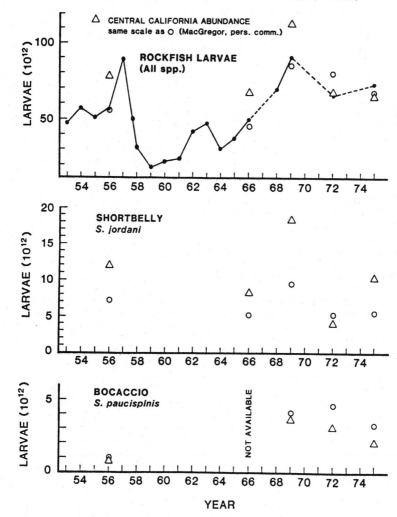

Figure 4.6. Abundances of larval rockfishes off California. Abundance indexes were provided by J. MacGregor, Southwest Fisheries Center, La Jolla, CA.

46

lalandei) is a popular target of southern California
recreational fishermen who take mainly the northward
migrants that reach the SCB. Yellowtail appeared to
have increased in abundance since the end of the
commercial fishery in the 1950s (MacCall et al., 1976),
but have recently declined due to several years of poor
recruitment (Crooke, 1983). Intense fisheries for
thresher sharks (_Alopias vulpinus_) and bonito sharks
(_Isurus oxyrinchus_) have developed recently (Cailliet
and Bedford, 1983), and may be expected to reduce
rapidly the abundance of those low fecundity species
(cf. Holden, 1977). The pre-fishery abundance of
predatory fishes is difficult to estimate, but should
be severalfold larger than their peak historical
harvests, perhaps by one or two hundred thousand tons.
Their present-day abundance is probably 10 to 25
percent of the original level, and the corresponding
reduction in prey consumed should have had a beneficial
effect on net productivity of the latter species.

Many other species play important roles in the
California Current ecosystem. For example, market
squid (_Loligo opalescens_) supports a commercial
fishery, and is a major component of many predators'
diets. Unfortunately, little is known about the
abundances and variability of these species.

MULTISPECIES CONSIDERATIONS

Our knowledge of multispecies interactions is
surprisingly limited in view of our extensive knowledge
of the California Current. As was suggested in the
preceding discussion, it is difficult to assess the
impact of the loss of the sardine from the ecosystem.
Both natural variability and fisheries have affected
other components of the ecosystem, and responses are
often indirect and confounded. Our knowledge has been
gained mostly since the beginning of the CalCOFI Program in 1950, a period in which the sardine has had
little biological influence. While it seems reasonable
to assume that loss of the sardine has had a significant impact on the ecosystem, the alternative hypothesis should also be considered: The ecosystem may be
well adapted to a highly fluctuating sardine abundance,
and its disappearance may have had relatively few
identifiable effects.

In a few cases, predator requirements have been
defined quantitatively. The brown pelican (_Pelecanus
occidentalis californicus_) is a designated "endangered
species" due to pesticide contamination and associated
reproductive failure about 1970, a condition from which
the California colonies have largely recovered (Anderson et al., 1975; Anderson and Gress, 1983). The relationship between pelican breeding success and anchovy

abundance (its main prey) is well-established (Figure 4.7, also see Anderson et al., 1980, 1982). If sardines once again become abundant, brown pelicans are likely to include them in their diet and should enjoy higher breeding success and less dependence on the anchovy (cf. Crawford and Shelton, 1978). In this context it is relevant that brown pelicans ceased breeding effectively at Pt. Lobos, near Monterey, at the same time that the Pacific sardine fishery failed (MacCall, in press). The relationship between brown pelican breeding success and anchovy abundance is recognized explicitly in the Northern Anchovy Fishery Management Plan developed by the Pacific Fishery Management Council (MacCall et al., 1983). Direct relationships between seabird reproduction and anchovy abundance have been shown also for western gulls (Larus occidentalis) and Xantu's murrelets (Endomychuura hypoleuca) by Hunt and Butler (1980), and for elegant terns (Sterna elegans) by Schaffner (1982). Despite their relatively high metabolic requirement, seabirds probably are not abundant enough in the California Current to have a major long-term impact on abundance of prey fishes. However, at times, seabird impact on forage may be large. For example, Wiens and Scott (1975) estimate that the millions of sooty shearwaters (Puffinus griseus) passing along the Oregon coast during their late summer southward migration may consume over 450 tons of anchovies per day, and larger quantities may be assumed for the much longer California coastline.

Pinnipeds consume large quantities of fish, and are increasing at a rapid rate in California. The approximately 70,000 California sea lions have been estimated to consume 100,000 to 300,000 tons of fish, and the 40,000 elephant seals may annually consume 400,000 to 1,000,000 tons, mostly of demersal fish (DeMaster, 1983). Daily rations are well known for pinnipeds, due to experience with captive animals, but the response to fluctuations in abundance and species composition of forage in the ocean is poorly known. Per capita reproductive rates of pinnipeds appear relatively constant except for severe declines during oceanic warmings such as 1982-1983.

A particularly difficult issue is the relationship between anchovy abundance and catch rates of large predatory fish. California's recreational fishermen strongly feel that abundant forage is necessary to attract large fishes to the region. Anchovies are certainly important to recreational fishing operations, being used as live bait, and anchovy schools provide visual clues to the location of predator fishes. The importance or value of these uses of anchovy is very difficult to quantify, but recognition of the large

48

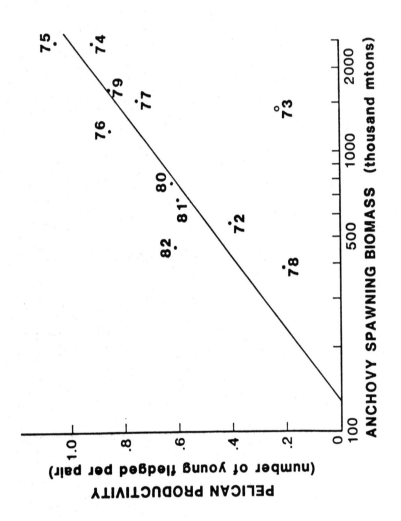

Figure 4.7. Relation between brown pelican productivity (Anderson et al., 1980, 1982) and anchovy spawning biomass (from MacCall et al., 1983). The 1981 and 1982 productivity values are unpublished data.

recreational fishing constituency in California has influenced anchovy harvesting policy, and constrained U.S. commercial fishery development in the early 1970s. More recently, the Pacific Fishery Management Council has chosen to pursue a harvest policy which seeks a moderate, relatively constant allowable anchovy harvest, and establishes a minimum level of anchovy abundance below which non-bait uses are curtailed (MacCall et al., 1983).

Multispecies considerations are operationally important to management of many species in California, although the biological basis of management remains single-species models. Incidental catches of non-target species are regulated, especially in the case of depleted stocks such as sardine and Pacific Ocean perch (Sebastes alutus). In order to prolong fishing opportunities, some fishing seasons have been established interactively. For example, in California, the anchovy reduction (fish meal) fishery ends on June 30, when summer live bait demand is high. The Pacific mackerel fishery opens on July 1, preserving continuity of opportunity for the California purse seine fleet (Klingbeil, 1983).

OUTLOOK

Developments in the California Current ecosystem and its management during the remaining years of the 20th century should be very interesting. The Pacific sardine recently has shown signs of increase (Bedford et al., 1983), although the resource presently remains severely depleted in the California Current. Recovery of depleted coastal pelagic resources often occurs suddenly, with production of a large year class. Examples are the strong 1976 and 1978 year classes of Pacific mackerel in California (Klingbeil, 1983) and the recent increase in sardines off Japan (Kondo, 1980). If recovery of the sardine is rapid, ecosystem effects may be somewhat easier to assess than they would be for gradual changes.

The major U.S. anchovy fishery produces fish meal, a high protein animal feed supplement which has declined in demand due to replacement by enriched soybean meal (Thomson, 1984). The Mexican fishery has grown since the mid-1970s, and Mexican domestic demand for fish meal is expected to remain strong. Consequently, the major harvest of anchovy now occurs south of the U.S.-Mexican border, where harvests are unregulated.

Pinniped populations are increasing rapidly, with some species increasing at over 10 percent per year, and interactions with fishermen are increasing disproportionately. Marine mammals are presently protected by the U.S. government under the Marine Mammal

50

Protection Act, but return of management to the individual states is being promoted. The debate between the preservationists and the fishermen is highly emotional, and the future course of marine mammal management in the California Current should be interesting to watch.

The greatest difficulty facing ecosystem management is the multitude of jurisdictions, often overlapping and conflicting, and each with its own objectives, responsibilities, and responsiveness. The portion of the California Current south of the U.S.-Mexican border is under Mexican jurisdiction, where some fauna such as marine mammals are protected, but fishery regulation is minimal. Access to Mexican waters by U.S. commercial fishermen has been severely restricted since 1982, which should have reduced fishing pressure on many species off Baja California. In California some species are managed by the federal government through the Pacific Fishery Management Council, while others are managed by state institutions, including the California legislature (most commercial fisheries) and the California Fish and Game Commission (recreational fisheries and some commercial fisheries). Marine mammals are managed by the U.S. Department of Commerce (and may be returned to state control), but seabirds are under the authority of the U.S. Department of the Interior. Despite this seeming confusion, coordination has been moderately successful when pursued on a single-species basis. Actual ecosystem management, if it were technically feasible, would be difficult to achieve unless management authority were better consolidated, with provisions for local representation and responsiveness. Such a consolidation appears unlikely in the foreseeable future.

REFERENCES

Ahlstrom, E., Moser, H. G., and Sandknop, E. 1978. Distributional atlas of fish larvae in the California Current region: Rockfishes, Sebastes spp., 1950 through 1975. Calif. Coop. Ocean. Fish. Invest. Atlas 26:178 pp.

Ainley, D., and Lewis, T. 1974. The history of Farallon Island marine bird population, 1854-1972. Condor 76:432-446.

Anderson, D., and Gress, F. 1983. Status of a northern population of California brown pelicans. Condor 85:79-88.

Anderson, D., Gress, F., and Mais, K. 1982. Brown pelicans: Influence of food supply on reproduction. Oikos 39:23-31.

Anderson, D., Gress, F., Mais, K., and Kelly, P. 1980. Brown pelicans as anchovy stock indicators and their relationships to commercial fishing. Calif. Coop. Ocean. Fish. Invest. Rep. 21:54-61.

Anderson, D., Jehl, J., Risebrough, R., Woods, L., Deweese, L., and Edgecomb, W. 1975. Brown pelicans: Improved reproduction off the southern California coast. Science 190:806-808.

Anonymous. 1977. California Current ecosystem. In Mammals in the seas. pp. 235-243. Ed. by S. Holt. FAO ACMRR working party on marine mammals. Fish. Ser. 1(5):264 pp.

Antonelis, G. A., Jr., and Fiscus, C. H. 1980. The pinnipeds of the California Current. Calif. Coop. Ocean. Fish. Invest. Rep. 21:68-78.

Bedford, D., Jow, T., Klingbeil, R., Read, R., Spratt, J., and Warner, R. 1983. Review of some California fisheries. Calif. Coop. Ocean. Fish. Invest. Rep. 24:6-10.

Cailliet, G., and Bedford, D. 1983. The biology of three pelagic sharks from California waters, and their emerging fisheries: a review. Calif. Coop. Ocean. Fish. Invest. Rep. 24:57-69.

Collins, R. 1981. Pacific coast croaker resources. Mar. Rec. Fish. 6:41-49.

Collins, R., Huppert, D., MacCall, A., Radovich, J., and Stauffer, G. 1980. Pacific bonito management information document. Calif. Dep. Fish Game, Mar. Resources Tech. Rep. 44:94 pp.

Collins, R. A., and MacCall, A. D. 1977. California's Pacific bonito resource, its status and management. Calif. Dep. Fish Game Mar. Res. Tech. Rep. 35:39 pp.

Crawford, R., and Shelton, P. 1978. Pelagic fish and seabird interrelationships off the coasts of South West and South Africa. Biol. Conserv. 14:85-109.

Crooke, S. 1983. Yellowtail, Seriola lalandei Valenciennes. Calif. Coop. Ocean. Fish. Invest. Rep. 24:84-87.

Daan, N. 1980. A review of replacement of depleted stocks by other species and the mechanisms underlying such replacement. Rapp. P.-v. Réun. Cons. int. Explor. Mer 177:405-421.

DeMaster, D. 1983. Annual consumption of northern elephant seals and California sea lions in the California Current (abstract). Calif. Coop. Ocean. Fish. Invest., Annual Conference 1983, Program and Abstracts.

Douglas, A. 1980. Geophysical estimates of sea-surface temperatures off western North America since 1671. Calif. Coop. Ocean. Fish. Invest. Rep. 21:102-112.

52

Hanan, D. 1983. Review and analysis of the bluefin
    tuna fishery in the eastern North Pacific Ocean.
    Fish. Bull., U.S. 81:107-119.
Hickey, B. 1979. The California Current system--
    Hypotheses and facts. Prog. in Oceanography 8:191-
    279.
Holden, M. 1977. Elasmobranchs. In Fish population
    dynamics, pp. 187-216. Ed. by J. Gulland. John
    Wiley and Sons, New York. 372 pp.
Hubbs, C. 1948. Changes in the fish fauna of western
    North America correlated with changes in ocean
    temperature. J. Mar. Res. 7:459-482.
Hunt, G., and Butler, J. 1980. Reproductive ecology
    of western gulls and Xantu's murrelets with
    respect to food resources in the Southern
    California Bight. Calif. Coop. Ocean. Fish.
    Invest. Rep. 21:62-67.
Klingbeil, R. 1983. Pacific mackerel: A resurgent
    resource and fishery of the California Current.
    Calif. Coop. Ocean. Fish. Invest. Rep. 24:35-45.
Kondo, K. 1980. The recovery of the Japanese
    sardine--The biological basis of stock-size
    fluctuations. Rapp. P.-v. Réun. Cons. int.
    Explor. Mer 177:332-354.
Lasker, R., and MacCall, A. 1983. New ideas on the
    fluctuations of the clupeoid stocks off Cali-
    fornia. In CNC/SCOR Proceedings of the Joint
    Oceanographic Assembly 1982--General Symposia, pp.
    110-120. Ottawa. 189 pp.
Laurs, R. 1983. The North Pacific albacore--An
    important visitor to California Current waters.
    Calif. Coop. Ocean. Fish. Invest. Rep. 24:99-106.
Lenarz, W. 1980. Shortbelly rockfish, Sebastes
    jordani: A large unfished resource in waters off
    California. Mar. Fish. Rev. 42(3-4):34-40.
MacCall, A. 1979. Population estimates for the waning
    years of the Pacific sardine fishery. Calif.
    Coop. Ocean. Fish. Invest. Rep. 20:72-82.
MacCall, A. 1983. Variability of pelagic fish stocks
    off California. In Proceedings of the expert
    consultation to examine changes in abundance and
    species composition of neritic fish resources.
    San Jose, Costa Rica, 18-29 April 1983, pp. 101-
    112. Ed. by G. Sharp and J. Csirke. FAO Fish.
    Rep. 291(2):1-553.
MacCall, A. In Press. Seabird-fishery trophic inter-
    actions in eastern Pacific boundary currents:
    California and Peru. In Marine birds: Their
    feeding ecology and commercial fisheries relation-
    ships. Ed. by D. Nettleship, G. Sanger, and P.
    Springer. Spec. Publ., Can. Wildl. Serv., Ottawa.
MacCall, A., and Methot, R. 1983. The historical
    spawning biomass estimates and population model in

the 1983 anchovy fishery management plan. Southw. Fish. Center Admin. Rep. LJ-83-17. 53 pp.

MacCall, A., Methot, R., Huppert, D., and Klingbeil, R. 1983. Northern anchovy fishery management plan. October 24, 1983. Pacif. Fish. Management Council, 526 S.W. Mill St., Portland OR 97201.

MacCall, A., and Stauffer, G. 1983. Biology and fishery potential of jack mackerel (Trachurus symmetricus). Calif. Coop. Ocean. Fish. Invest. Rep. 24:46-56.

MacCall, A., Stauffer, G., and Troadec, J. P. 1976. Southern California recreational and commercial marine fisheries. Mar. Fish. Rev. 38(1):1-32.

McEvoy, A. 1979. Economy, law, and ecology in the California fisheries to 1925. Dissertation, Univ. California, San Diego. 484 pp.

Methot, R. 1983. Management of California's nearshore fishes. Mar. Rec. Fish. 8:161-172.

Miller, D., and Lea, R. 1972. Guide to the coastal marine fishes of California. Calif. Dep. Fish Game, Fish Bull. 157:235 pp.

Morejohn, G. 1976. Evidence of survival to recent times of the extinct flightless duck, Chendytes lawi (abstract). Pacif. Seabird Group Bull. 3(1):31-32.

Murphy, G. 1966. Population biology of the Pacific sardine (Sardinops caerulea). Proc. Calif. Acad. Sci. 4th Ser. 34(1):1-84.

Odum, E. 1971. Fundamentals of ecology, 3rd edition. W. B. Saunders, Philadelphia. 226 pp.

Parrish, R., Bakun, A., Husby, D., and Nelson, C. 1983. Comparative climatology of selected environmental processes in relation to eastern boundary current pelagic fish reproduction. In Proceedings of the expert consultation to examine changes in abundance and species composition of neritic fish resources. San Jose, Costa Rica, 18-29 April, 1983, pp. 731-788. Ed. by G. Sharp and J. Csirke. FAO Fish. Rep. 291(3):557-1224.

Parrish, R., and MacCall, A. 1978. Climatic variation and exploitation in the Pacific mackerel fishery. Calif. Dep. Fish Game, Fish Bull. 167:110 pp.

Radovich, J. 1961. Relationships of some marine organisms of the northeast Pacific to water temperatures particularly during 1957 through 1959. Calif. Dep. Fish Game, Fish Bull. 112:62 pp.

Reid, J., Roden, G., and Wyllie, J. 1958. Studies of the California Current system. Calif. Coop. Ocean. Fish. Invest. Rep., 1 July 1956 to 1 January 1958:27-56.

Santander, H., Alheit, J., MacCall, A., and Alamo, A. 1983. Egg mortality of the Peruvian anchovy (Engraulis ringens) caused by cannibalism and predation by sardines (Sardinops sagax). In Proceedings of the expert consultation to examine changes in abundance and species composition of neritic fish resources. San Jose, Costa Rica, 18-29 April, 1983, pp. 1011-1025. Ed. by G. Sharp and J. Csirke. FAO Fish. Rep. 291(3):557-1224.

Schaffner, F. 1982. Aspects of the reproductive ecology of the elegant tern (Sterna elegans) at San Diego Bay. M.S. Thesis, San Diego State Univ., San Diego, Calif. 185 pp.

Schultze, D. 1983. California barracuda life history, fisheries and management. Calif. Coop. Ocean. Fish. Invest. Rep. 24:88-96.

Smith, P. 1972. The increase in spawning biomass of northern anchovy, Engraulis mordax. Fish. Bull., U.S. 70:849-874.

Smith, P. 1978. Biological effects of ocean variability: Time and space scales of biological response. Rapp. P.-v. Réun. Cons. int. Explor. Mer 173:117-127.

Soutar, A., and Isaacs, J. 1969. History of fish populations inferred from fish scales in anaerobic sediments off California. Calif. Coop. Ocean. Fish. Invest. Rep. 13:63-70.

Soutar, A., and Isaacs, J. 1974. Abundance of pelagic fish during the 19th and 20th centuries as recorded in anaerobic sediment off the Californias. Fish. Bull., U.S. 72:257-273.

Sowls, A., DeGange, A., Nelson, J., and Lester, G. 1980. Catalog of California seabird colonies. U.S. Dep. Int., Fish Wildl. Serv., Biol. Serv. Progr. FWS/OBS 37/80 371 p.

Thomson, C. 1984. A model of fishmeal supply and demand in the United States 1966-1981. Southw. Fish. Center Admin. Rep. LJ-84-04. 27 pp.

Vojkovich, M., and Reed, R. 1983. White seabass, Atractoscion nobilis, in California-Mexican waters: Status of the fishery. Calif. Coop. Ocean. Fish. Invest. Rep. 24:79-83.

Walker, P., and Craig, S. 1979. Archaeological evidence concerning the prehistoric occurrence of sea mammals at Point Bennett, San Miguel Island. Calif. Fish Game 65:50-54.

Warter, S. 1978. The extinct flightless seaducks of southern California (abstract). Pacif. Seabird Group Bull. 5(2):89.

Wiens, J., and Scott, J. 1975. Model estimation of energy flow in Oregon coastal seabird populations. Condor 77:439-452.

Michael P. Sissenwine

5. Perturbation of a Predator-Controlled Continental Shelf Ecosystem

ABSTRACT

The continental shelf area off the northeastern USA is highly productive at all trophic levels. During the late 1950s, total catch was about 400,000 tons annually. The catch increased in the 1960s and reached a peak of 1.2 million tons in 1972. The dramatic increase in catch resulted from a sixfold increase in fishing effort, primarily by distant water fishing vessels from Europe and Japan. The catch declined rapidly during the mid- and late 1970s due to depletion of the fishery resource and international and domestic constraints on fishing.

The fish and squid community declined in biomass by about 60% from the mid-1960s to the mid-1970s. There has been only a modest recovery since some of the fishing pressure was relieved. The decline in abundance would have been more severe if some of the depleted species (e.g., herring, haddock) were not at least partially compensated for by increases in other species (e.g., squid, dogfish, sand lance).

An energy budget was developed (and revised) in order to better understand the response of the ecosystem to perturbation by overfishing. Production of the fishery resource was partitioned between 13 species or species groups. Since relatively little is known about pre-exploitable fish (i.e., individuals which are too small and too young to be captured by commercial or research vessel survey gear) a model was used to estimate their production and consumption based on the initial number and weight of individuals in the cohort and their number and weight when they reach exploitable size.

While pre-exploitable fish are only 10% of the biomass of exploitable fish, their consumption is nearly as great and their production is two and a half times as high. Fish consume most of their own produc-

55

tion (61-93%). Other consumers are marine mammals, birds, large pelagic migratory fish (e.g., sharks) and humans. The response of the fishery resource to the perturbation of overfishing may have been moderated due to predation.

INTRODUCTION

The purpose of the symposium on Variability and Management of Large Marine Ecosystems is to exchange information, so that we can learn from geographic similarities and differences, in order to enhance management. This chapter contributes to the goal by characterizing the large marine ecosystem of the northeast continental shelf of the USA.

The chapter is divided into four sections. Following the Introduction, a brief description of the oceanography (i.e., topography, hydrography, and productivity) of the region is given. The next section describes the major fisheries and draws preliminary conclusions about the system based on its response to the extreme perturbation of overfishing. The final section examines the ecology of the system in more detail by summarizing the results of much research into an energy budget. The energy budget has important implications relative to the stability of the system, its robustness to fishing, and future research.

OCEANOGRAPHY

The area of the continental shelf to 200-m depth off the northeastern USA (Cape Hatteras, North Carolina, to Nova Scotia, Canada) is about 260,000 square kilometers. It is depicted in Figure 5.1. The topography is diverse with the deep basins of the Gulf of Maine, the broad continental shelf of the Mid-Atlantic Bight, the shoals of Georges Bank, and the submarine canyons of the continental slope. The steepness of the sea bed is indicated by isobaths at 20-m intervals (Figure 5.2).

The Mid-Atlantic area is influenced by three major estuaries--Chesapeake Bay, Delaware Bay, and Hudson-Raritan. The seaward extent of estuarine plumes is highly variable, but may extend 70 kilometers.

The total freshwater discharge to the Gulf of Maine is comparable to the Mid-Atlantic area, but it is distributed among a large number of small rivers. As a result, estuarine plumes are not observed; instead there is a narrow coastal band of low salinity.

Bumpus (1973), Hopkins and Garfield (1977), and Ingham (1982) describe the oceanography of the region. The average circulation (Figure 5.3) involves seasonally variable gyres in the Gulf of Maine and on Georges

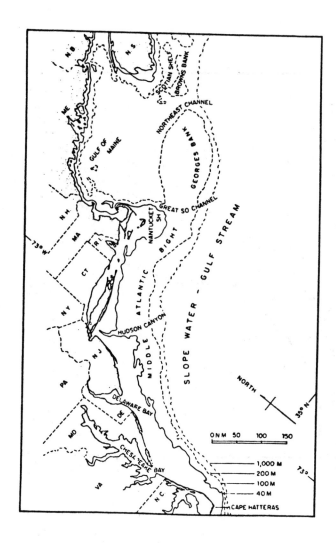

57

Figure 5.1.   The continental shelf area off the
northeastern coast of the USA; Cape Hatteras, North
Carolina, to Nova Scotia, Canada (Ingham, 1982).

58

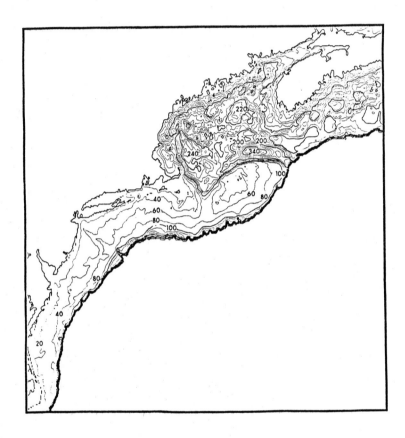

Figure 5.2.   Bathymetry of the continental shelf to 200-m depth (Cape Hatteras, N.C., to Nova Scotia, Canada).

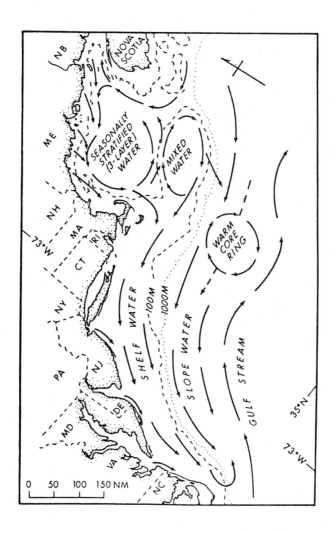

Figure 5.3. General surface layer circulation of Northwestern Atlantic coastal and offshore waters (Ingham, 1982).

Bank, and a slow westward and southwestward flow on the continental shelf and the Mid-Atlantic Bight, and a southwestward flow in the slope water.

The gyre in the Gulf of Maine is cyclonic (counterclockwise). It is strongest in the spring and early summer months. During fall and winter the southern portion may break down and allow water to drift onto Georges Bank and southward. The central portion of the Gulf of Maine is seasonally stratified. A temperature range of the surface water is from 1° to 15°C. The bottom water is saltier and remains at 6° to 8°C year-round.

The circulation of Georges Bank also involves a seasonably variable gyre, but it is anticyclonic (clockwise). The gyre is well developed in spring and summer, but by fall and winter its west side breaks down and there is drift to the Mid-Atlantic area. Because of strong tidal currents, the water over the center of Georges Bank is well mixed throughout the year.

The shelf water in the Mid-Atlantic area undergoes stratification in spring and summer. Coastal areas experience episodes of low dissolved oxygen. Beneath the seasonal thermocline and over the shelf south of Cape Cod there is a layer of water called the cold pool. During winter and early spring the water column is isothermal with the coldest water found near shore with a weak horizontal gradient towards warmer water offshore. There is an abrupt gradient in salinity and temperature at the edge of the continental shelf.

The region is highly productive. Figure 5.4 summarizes the annual primary productivity in grams carbon per meter squared per year (O'Reilly and Busch, 1984). Phytoplankton production was estimated using the ^{14}C simulated in situ sunlight method during 23 surveys, at 628 stations, during 1977-1980 throughout the year. Particulate production accounts for 82-90% of the total organic carbon production reported in Figure 5.4. The northeastern USA is 2-4 times more productive than most continental shelf ecosystems (e.g., North Sea, Scotian Shelf, Baltic Sea, coastal Japan; see O'Reilly and Busch, 1984; Sherman et al., in press).

The high level of primary productivity of the region is at least in part related to its hydrography. The isothermal conditions on Georges Bank allow for the rapid return of nutrients from the bottom to the euphotic zone. Primary productivity in the Mid-Atlantic area is enhanced by nutrients from the estuaries, in some cases due to anthropogenic activity. Although primary production in the Gulf of Maine is lower than elsewhere in the region, it is still relatively high

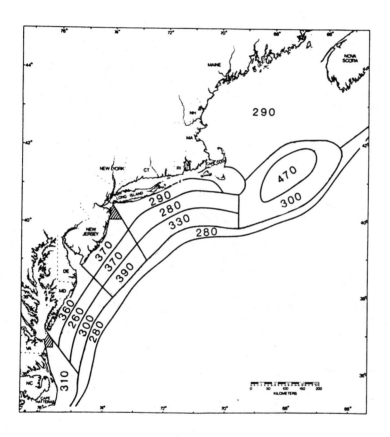

Figure 5.4.   Estimates of annual phytoplankton production (particulate and dissolved organic carbon) by region, in grams carbon per meter squared per year. Data collected in hatched area not included in the regional estimate of annual production.   After O'Reilly and Busch (1984).

compared to other continental shelf areas. A high level of primary productivity translates into productive fisheries on a per unit area basis. These are described in the next section.

FISHERY

Prior to 1960, the fishery resources off the northeastern USA were exploited primarily by domestic and secondarily by Canadian vessels. The target species were haddock, cod, redfish, and several species of flounder. Sea herring were also important, but they were only exploited in the coastal waters of the State of Maine. During the late 1950s, the total catch was about 400,000 tons annually. The catch increased in the 1960s and reached a peak of 1.2 million tons in 1972. This was followed by a decrease in total catch to 360,000 tons in 1978, with little change since then (Figure 5.5).

The dramatic increase in catch was caused by the entry of distant water factory fishing vessels from Europe and Japan. Standardized fishing effort (taking account of the greater fishing power of large distant water vessels) increased by sixfold from 1961 to 1972 (Figure 5.6; Clark and Brown, 1979). Virtually all the increase was due to the distant water vessels. During the early 1970s the International Commission for the Northwest Atlantic Fisheries (ICNAF) adopted regulations which resulted in a modest decrease in fishing effort. The era of intense exploitation by distant water fleets ended in 1977 when the USA extended its fisheries jurisdiction to 200 miles (including the entire continental shelf off the northeastern USA). Nevertheless, standardized fishing effort remained high relative to the period prior to the arrival of the distant water fleets, due to an increase in the number of U.S. vessels and an increase in their fishing power.

The distant water fleet was less selective than the domestic and Canadian fishing vessels which depended on a relatively small number of species. Virtually the entire nekton (i.e., fish and squid) were subject to fishing, although the distant water fleets concentrated their effort on the dominant year classes of the most abundant species. Initially, herring were the target. Prior to 1961, the Georges Bank herring population was virtually unexploited. Due to a series of strong year classes during the late 1950s and early 1960s the population grew until it reached a peak in 1968. The catch also peaked in 1968 and then declined sharply. Fishing ceased in 1977. The Georges Bank population has been virtually extinct since then (Figure 5.7).

63

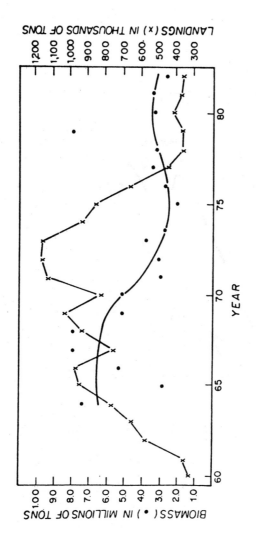

Figure 5.5. Annual catch (excluding menhaden and large
pelagic species, e.g., sharks and tuna) and estimated
biomass of "exploitable" fish and squid off the northeastern USA, Cape Hatteras to Nova Scotia.

64

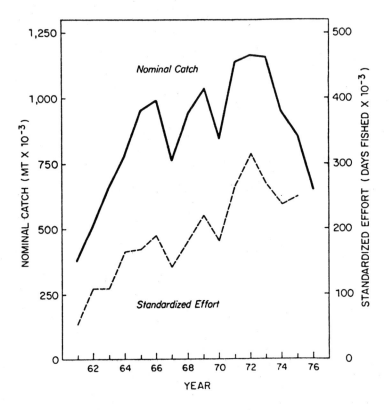

Figure 5.6. Annual catch (excluding menhaden and large pelagic species, e.g., sharks and tuna) and standardized fishing effort. After Clark and Brown (1977).

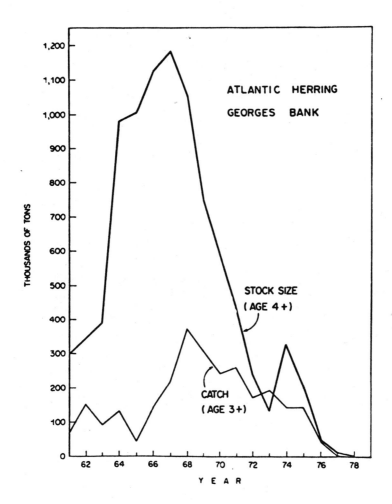

Figure 5.7.   Annual catches (age 3 and older) and
estimates of stock size (age 4 and older) of Atlantic
herring from Georges Bank area.   After Anderson (1984).

The distant water fleets diverted their effort from herring to Georges Bank haddock in 1965 and 1966 in order to take advantage of the very strong 1963 year class (Figure 5.8). The catch peaked in 1965 at about three times the long-term average. The domestic share of the catch was virtually unchanged. The fishery collapsed by 1970. There was a short-lived recovery in the late 1970s and early 1980s.

The distant water fleets also took advantage of a nearly virgin population of silver hake on Georges Bank (Figure 5.9). When the Georges Bank population dwindled, they turned their attention to silver hake in southern New England and the Mid-Atlantic area (Figure 5.10) and to other demersal species, such as cod and flounder.

Harvesting of pelagic species remained important throughout the era of the distant water fleets. When the Georges Bank herring population began to decline, the distant water fleet switched to mackerel, taking advantage of the outstanding 1967 year class (Figure 5.11). Later, during the 1970s, effort shifted to squid.

How did the fish and squid community as a whole respond to the dramatic increase in harvest and fishing effort (presumably a sixfold increase in fishing mortality)? Clark and Brown (1977, 1979) used research vessel bottom trawl survey data and commercial fishery statistics to estimate the biomass of the exploitable component of the fish and squid community. The term "exploitable" means those individuals which are represented by the aforementioned types of data, generally larger than 15 cm in length in one year of age.

Estimates of biomass, based on Clark and Brown's approach, are given in Figure 5.5. A trend line is drawn in an attempt to smooth the data. The 1965 and 1979 estimates of biomass are de-emphasized due to ancillary information which indicates that pelagic species were under- and over-represented, respectively.

The response of the fish and squid community to an extreme perturbation of fishing was a 60% decline in biomass from the mid-1960s to the mid-1970s. There has been a modest recovery since some of the fishing pressure was relieved due to ICNAF regulations during the mid-1970s and USA extended jurisdiction since 1977.

Although the decline in fish and squid biomass was substantial, there appears to be some degree of stability in spite of the extreme perturbation. Some species were virtually eliminated (e.g., Georges Bank herring, Georges Bank haddock), but these losses were partially compensated for by increases in other species (e.g., squid). The decline in silver hake, a predatory species, was partially offset by an increase in abundance

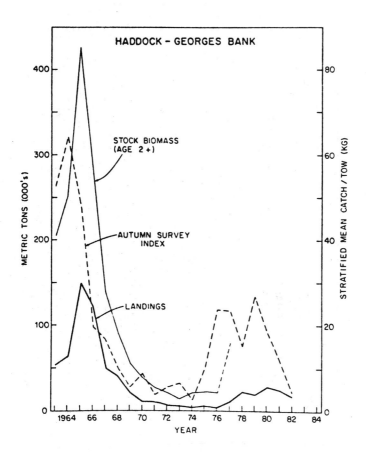

Figure 5.8. Annual landings, stock biomass (age 2 and older) and relative abundance as indicated by autumn bottom trawl surveys for Georges Bank haddock. After Anderson (1984).

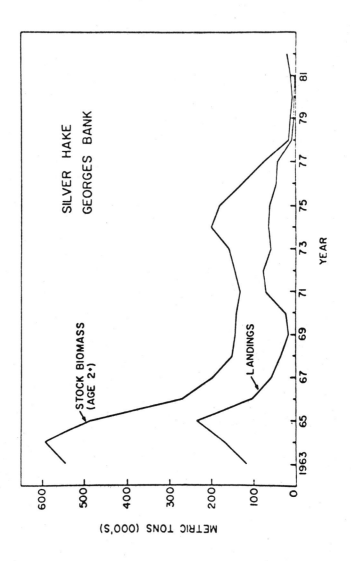

Figure 5.9. Annual landings and stock biomass (age 2 and older) for Georges Bank silver hake. After Anderson (1984).

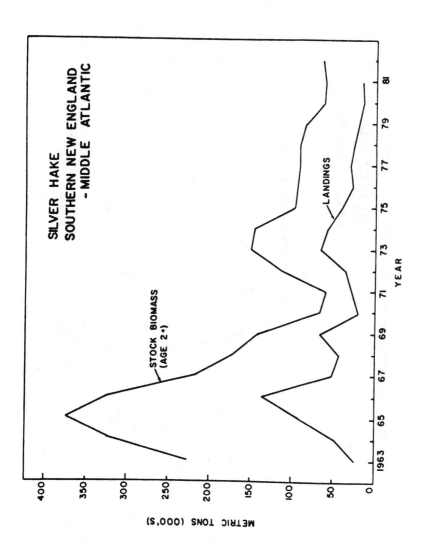

Figure 5.10. Annual landings and stock biomass (age 2 and older) for southern New England--Mid-Atlantic area silver hake. After Anderson (1984).

70

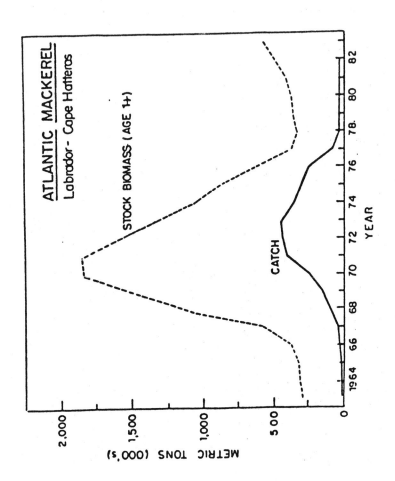

Figure 5.11. Annual catch and stock biomass (age 1 and older) of Atlantic mackerel from Cape Hatteras to Labrador. After Anderson (1984).

of predatory dogfish and bluefish (Figure 5.12). The decline in planktivorous herring and mackerel was partially offset by an increase in abundance of sand lance (Sherman et al., 1981), a species not taken account of by Clark and Brown's method. Ichthyoplankton surveys have shown a sharp increase in sand lance abundance since the mid-1970s (Morse, 1982; also Figure 5.8 of Sherman, this volume).

Figure 5.13 illustrates the switching characteristic of the fishery. The annual catch of 11 species or species groups and the total for Georges Bank is given for 1955 through 1978. The annual catch is highly variable for each species, but less so for the total; the variance of the total is indeed less than the sum of the variances of the individual species, although most of the difference is due to herring and mackerel which are the dominant species in the catch.

The response of the fishery resources to an extreme perturbation of overfishing offers a clue that there are stabilizing mechanisms which are operative at the community level. In order to identify the mechanisms it is necessary to understand how the species interact. A better understanding has been achieved by developing (and revising) an energy budget for a particularly well-studied large marine ecosystem off the northeastern USA--Georges Bank. The energy budget is described in the next section.

ENERGY BUDGET

Georges Bank and the surrounding region has been the focus of intense ecological studies (Grosslein et al., 1979; Fogarty et al., in press). These include research on phytoplankton, zooplankton, ichthyoplankton, benthos, and nekton (i.e., finfish and squid); ^{14}C method, bongo net plankton surveys, bottom trawl surveys, and benthic grab samples have been employed. The results of these investigations have been used to develop (and revise) an energy budget for Georges Bank (Cohen et al., 1982; Sissenwine et al., 1984; Cohen and Grosslein, in press). The description that follows (abstracted from Sissenwine, in press) focuses on production and consumption of fish.

Grosslein et al. (1980) estimated the exploitable biomass of 11 fish species or species groups and two squid species; yellowtail flounder, cod, haddock, silver hake, mackerel, herring, redfish, red hake, pollock, flounder (other than yellowtail), other finfish, short-finned squid, and long-finned squid. Biomass was expressed in kcal/m^2 (assuming 1 Kcal/g wet weight). Average values for three-year periods (1964-1966 and 1973-1975) were reported (Figure 5.14). The

72

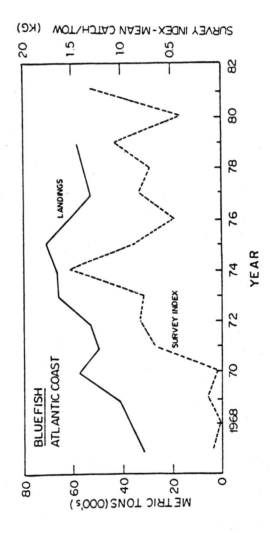

Figure 5.12. Annual landings and relative abundance is indicated by research vessel bottom trawl surveys for bluefish off northeastern USA from Cape Hatteras to Nova Scotia. After Anderson (1984).

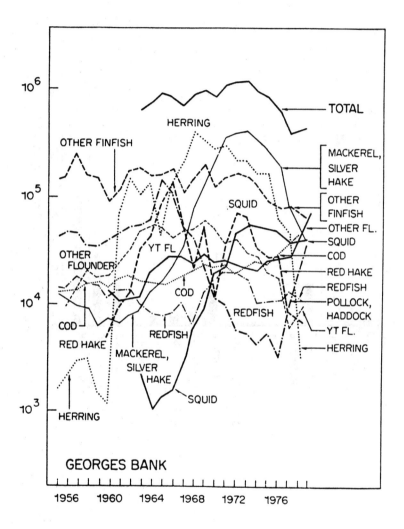

Figure 5.13.   Annual catch (total and by species) for
Georges Bank.   After Hennemuth (1979).

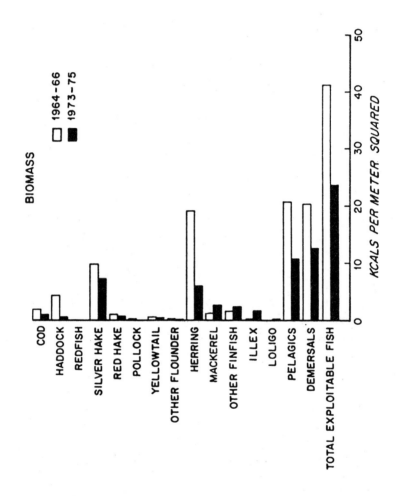

Figure 5.14. Biomass of exploitable fish on Georges Bank for 1964-1966 and 1973-1975. After Grosslein et al. (1980).

former period was one of high abundance when distant
water fishing fleets were still increasing. The latter
period was one of reduced abundance following a decade
of intense fishing.

The production and consumption rate of a fish is
dependent on its size. The size composition of the
population varies as a result of fishing and fluctuations in recruitment. Therefore, production to
biomass and consumption to biomass ratios also vary.
Grosslein et al. (1980) examined species and temporal
variability in these ratios. They calculated annual
estimates for the first six species listed above over a
10-year period (1963-1972). These species are a
reasonable cross section of the exploitable fish on
Georges Bank and they constitute most of the biomass
during the periods considered.

Annual production by each group was calculated as
the sum of somatic and gonadal growth based on the
literature (e.g., growth functions, fecundity functions, maturity ogives). Annual assimilated consumption was calculated by adding the energy expended from
metabolism to production. Metabolic rate was taken to
be twice the energy required for routine metabolism
based on a power function of body weight, fit to
laboratory data summarized from the literature by
Geoffrey Laurence (see Grosslein et al., 1980). The
assimilated consumption was assumed to be 80% of the
total. Production to biomass and consumption to
biomass ratios were calculated by dividing by biomass.

The overall mean production to biomass and consumption to biomass ratios (0.46, 4.10, respectively)
for yellowtail flounder, cod, haddock, silver hake,
mackerel, and herring were applied to the other species
or species groups, except for redfish and squid. A low
production to biomass ratio of 0.25 was assumed for
redfish, a slow-growing species. For squid, which grow
quickly, a high ratio of 1.5 was assumed. Similarly,
consumption to biomass ratios of 3.0 and 7.0 was
assumed for redfish and squid, respectively. These
species were only a minor portion of the biomass of
Georges Bank during the periods considered, so the
overall analysis is not sensitive. The resulting
estimates of production and consumption by the
exploitable fish of Georges Bank are given in Figures
5.15 and 5.16.

Relatively little is known about the population
dynamics of the young (pre-exploitable) fish, particularly after the larval stage and before they grow
large enough to be caught in trawls. However, some
valuable information is available. The initial number
and biomass of the cohort of young fish can be estimated from the abundance of adults, the proportion of
total adult production used for reproduction, and

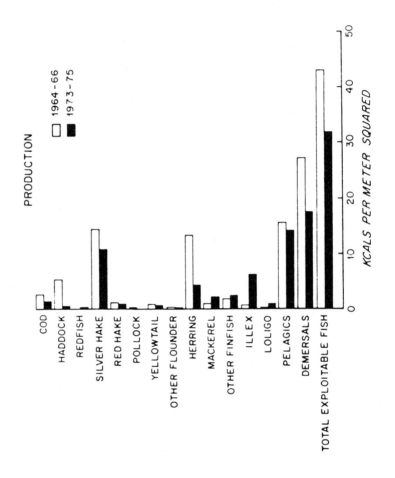

Figure 5.15. Production of exploitable fish on Georges Bank for 1965-1966 and 1973-1975. After Grosslein et al. (1980).

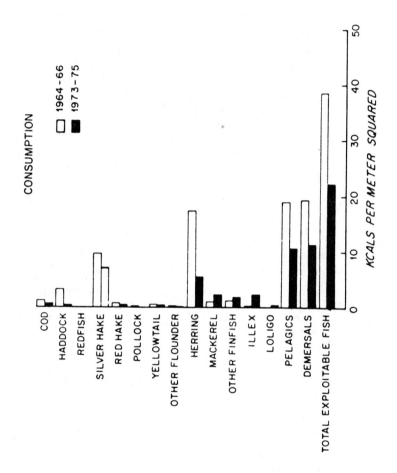

Figure 5.16. Consumption of exploitable fish on Georges Bank for 1964-1966 and 1973-1975. After Grosslein et al. (1980).

average weight of an egg. As the young fish reach
exploitable size (recruitment) the number and biomass
can be estimated from trawl survey data and fishery
statistics. With these beginning and end points known
for the pre-exploitable fish, a simple model can be
used to calculate rough estimates of production and
consumption rate (Sissenwine et al. 1984). The model
assumes that the growth rate of individual fish and of
the entire group of pre-exploitable fish is exponential. Metabolic rate, production and consumption are
calculated using an approach analogous to the method
applied to exploitable fish.

The model is an oversimplification of reality. It
assumes a constant growth rate and ignores differences
between species. Therefore, the results were subjected
to a sensitivity analysis (Sissenwine, in press). The
conclusions drawn in this paper are robust.

The biomass, consumption, and production of fish
and squid on Georges Bank, for 1964-1966 and 1973-1975,
are partitioned between demersal, pelagic, and pre-
exploitables (Figure 5.17). During the former period,
exploitable fish biomass was nearly equally divided
between pelagic and demersal species. However, due
mainly to the dominance of herring among the pelagics,
demersal production was nearly twice pelagic production; herring have a relatively low production to
biomass ratio.

After a decade of heavy exploitation, exploitable
biomass had declined by 1972-1975 to 43% of the earlier
level. Consumption declined by about the same percentage. The biomass remained about equally divided
between pelagics and demersals, but in contrast with
the earlier period, the contribution of pelagics to
total production increased, becoming roughly equal to
demersal production. The explanation lies in the
partial replacement of herring by mackerel and squid.
Mackerel and squid have higher production to biomass
ratios. As a result, production declined by only 26%,
less than the percentage decline in biomass.

Pre-exploitable fish biomass was only 10% of
exploitable biomass. Yet these young fish consumed
nearly as much and produced two and a half times as
much as the exploitable fish. Since the model of
consumption, production, and biomass for young fish
does not consider species composition, there is no
basis for comparing their relative importance between
periods.

Grosslein et al. (1980) summarized stomach contents data for several important species. Herring and
redfish prey on planktonic crustaceans. Haddock,
yellowtail flounder, and other flounder prey on poly-
chaetes and benthic crustaceans. Cod, silver hake, and
several other species prey on fish. Using more detailed

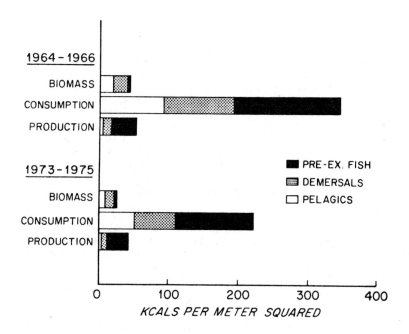

Figure 5.17.   Biomass, consumption, and production of
pelagic, demersal, and pre-exploitable fish on Georges
Bank for 1964-1966 and 1973-1975.   After Sissenwine et
al. (1984).

data, Cohen and Grosslein (in press) estimated that
cod, silver hake, pollock, mackerel, and the other
finfish category consumed 38%, 74%, 24%, 10%, 50%, and
30% fish, respectively. According to these percent-
ages, the finfish and squid community consumed 42.2 and
39.3 Kcal m^{-2} yr^{-1} of its own production during 1964-
1966 and 1973-1975, respectively.

The energy budget for Georges Bank is summarized
in Table 5.1. Estimates of particulate phytoplankton,
microzooplankton, macrozooplankton, and benthic production are from Cohen and Grosslein (in press).

Fish production was 1.3% to 2.1% of particulate
primary productivity. Considering the complexity of
the food web, trophic efficiency must be high relative
to traditional thinking (10%, Slobodkin, 1961). One
implication of the result is that the energy budget is
"tight," and fish production is ultimately limited by
their food resource. In fact, Table 5.1 indicates that
fish consume from 31% to 50% of the production of suitable prey types. This is remarkably high considering
that microzooplankton (about 60% of the zooplankton
production) is a suitable food for only a brief period
during the life cycle of fish, and that consumers other
than fish and squid are dependent on the components of
the ecosystem which have been labeled in Table 5.1 as
potential fish prey.

Fish consume most of their own production (61-
93%). Other consumers are marine mammals, birds, large
pelagic migratory fish (e.g., sharks), and humans.
Estimates of their consumption are also included in
Table 5.1. The estimated total consumption of fish
ranges from 83% to 130% of production for the two
periods considered. The deviations of these estimates
from 100% are considered within the level of precision
of the available data. The partitioning of fish production between the various predators is represented in
the pie diagram given in Figure 5.18.

Silver hake are responsible for most of the
consumption of fish, and they prey heavily on pre-
exploitable fish. In fact, consumption of silver hake
and other species is such a high fraction of the total
fish production that one or the other estimate is
likely to be in error. Sissenwine et al. (1984)
concluded that the estimated production of pre-
exploitable fish is the most likely source of error,
but it is unlikely to be underestimated by more than
about 40%. Of course, if production of pre-exploitable
fish is underestimated, then their consumption is also
underestimated. In fact, Sissenwine et al. (1984)
concluded that there is probably inadequate benthic and
macrozooplankton production to support substantially
more pre-exploitable fish production on Georges Bank.
Therefore, predation by fish must cause considerable

TABLE 5.1.

Components of Georges Bank energy budget. Production
estimates and fish consumption estimates are based on
Sissenwine et al. (1984). Bird, mammal, and large
pelagic consumption estimates are based on Powers (in
press), Scott et al. (1983), and Cohen and Grosslein
(in press), respectively. Human consumption corre-
sponds to the average catch, 1968-1982.

$$\text{Kcal } m^{-2} yr^{-1}$$

Production		
Phytoplankton (particulate)	3780	
Zooplankton	496	
Benthos	106	
Fish (exploitable)	13-17	
(pre-exploitable)	29-52	
(total)	42-69	(1.3-2.1% of Phytoplankton Production)
Potential Fish Prey	644-671	
Fish Consumption		
All Prey	197-344	(31-50% of Potential)
Of Fish	39-42	(93 and 61% of their own production)
Consumption of Fish		
By Fish	39-42	
By Birds	2.0	
By Mammals	5.4	
By Large Pelagics	2.0	
By Humans	6.1	
Total	54.6-57.6	130 and 83% of Fish Produc- tion, respec- tively)

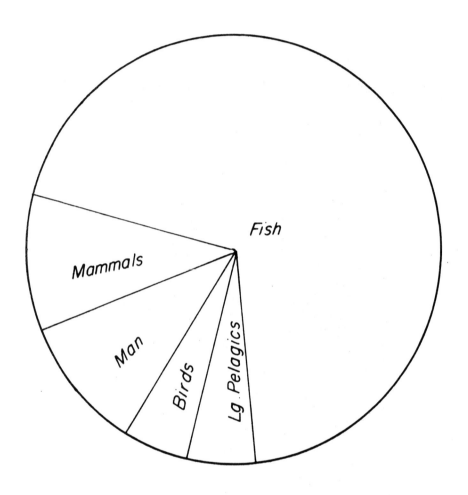

Figure 5.18. Proportion of total Georges Bank fish production consumed by fish, marine mammals, man, birds, and large pelagic (migratory) fish.

pre-exploitable mortality. Furthermore, there is empirical evidence (based on ichthyoplankton surveys, Sissenwine [1984]) that the mortality fraction (fraction dying during the life history stage) of post-larval pre-exploitable fish remains high, i.e., comparable to the mortality fraction of eggs and larvae.

Predation probably affects recruitment. If post-larval mortality is high, then there exists a potential that year-class strength is not established until this life stage since only a small variation in mortality would be necessary to account for a large change in recruitment.

There is clearly a biological basis for expecting compensation (a stabilizing mechanism) in production by the Georges Bank finfish community. They modify their own abundance by predation (cannibalism at the community level). It seems likely that this mechanism operates at a species as well as community level. Some species are cannibalistic (e.g., silver hake); furthermore, and abundance of fish as prey may enhance production of predators, ultimately resulting in compensation.

Cohen and Grosslein (in press) compared Georges Bank to other continental shelf ecosystems. These comparisons indicate that Georges Bank is not unique in its efficient conversion of primary productivity to fish production. For example, fish production is about 1% of particulate primary productivity for the North Sea, and about 3% for the East Bering Sea. These ecosystems are also "tight" like Georges Bank, and probably food limits their overall productivity as well. Furthermore, North Sea and the East Bering Sea fish consume a significant proportion of their own production (Laevastu and Larkins, 1981; Daan, 1983).

The fish community of the northeastern USA is predator controlled. As a result, its response to the severe perturbation of overfishing was probably moderated.

REFERENCES

Anderson, E. D. (Editor). 1984. Status of the fishery resources off the northeastern United States for 1983. NOAA Tech. Mem. NMFS-F/NEC-29, 132 pp.
Bumpus, D. F. 1973. A description of the circulation on the continental shelf of the east coast of the United States. Prog. Oceanogr. 6:111-157.
Clark, S. H., and Brown, B. E. 1977. Changes in biomass of finfishes and squid from Gulf of Maine to Cape Hatteras, 1963-1974, as determined from research vessel survey data. Fish. Bull, U.S. 75:1-21.

Clark, S. H., and Brown, B. E. 1979. Trends in biomass of finfish and squid in ICNAF subarea 5 and statistical area 6, 1964-1977, as determined from research vessel survey data. Investigacion Pesquera 43(1):107-122.

Cohen, E. B., and Grosslein, M. D. In press. Total ecosystem production on Georges Bank: Comparison with other marine ecosystems. In Georges Bank. Ed. by R. H. Backus. MIT Press, Cambridge.

Cohen, E. B., Grosslein, M. D., Sissenwine, M. P., Steimle, F., and Wright, W. R. 1982. An energy budget for Georges Bank. Can. Spec. Publ. Fish. Aquat. Sci. 59:95-107.

Daan, N. 1983. The ICES stomach sampling project in 1981: Aims, outline and some results. Northw. Atl. Fish. Organization Sci. Res. Doc. 83/IX/93.

Fogarty, M. J., Sissenwine, M. P., and Grosslein, M. D. In press. Fisheries research on Georges Bank. In Georges Bank. Ed. by R. H. Backus. MIT Press, Cambridge.

Grosslein, M. D., Brown, B. E., and Hennemuth, R. C. 1979. Research, assessment, and management of a marine ecosystem in the Northwest Atlantic: A case study. In Environmental biomonitoring, assessment, prediction, and management--Certain case studies and related quantitative issues. pp. 289-357. Ed. by M. I. Artuz, J. Cairns, Jr., G. P. Patil, and W. E. Waters. Int. Coop. Publ. House, Fairland, MD.

Grosslein, M. D., Langton, R. W., and Sissenwine, M. P. 1980. Recent fluctuations in pelagic fish stocks of the Northwest Atlantic, Georges Bank region, in relationship to species interaction. Rapp. P.-v. Réun. Cons. int. Explor. Mer 177:374-404.

Hennemuth, R. C. 1979. Man as predator. In Contemporary quantitative ecology and related economics. pp. 507-532. Ed. by G. P. Patil and M. Rosenzweig. Int. Coop. Publ. House, Fairland, MD.

Hopkins, T., and Garfield, N. 1977. Physical oceanography. Chapter 4. In Summary of environmental information--Continental shelf--Bay of Fundy to Cape Hatteras. U.S. Dep. Inter., Bur. Land Manage.

Ingham, M. C. (ed.). 1982. Summary of the physical oceanographic processes and features pertinent to pollution distribution in the coastal and offshore waters of the northeastern United States, Virginia to Maine. NOAA Tech. Mem. NMFS-F/NEC-17, 166 pp.

Laevastu, T., and Larkins, H. A. 1981. Marine fisheries ecosystem: Its quantitative evaluation and management. Fishing News Books, Ltd., Farnham, Surrey, England.

Morse, W. 1982. Spawning stock biomass estimates of sand lance, Ammodytes sp., off northeastern United States, determined from MARMAP plankton surveys 1974-1980. ICES C.M.1982/G:59.

O'Reilly, J. E., and Busch, D. A. 1984. Phytoplankton primary production on the northwestern Atlantic shelf. Rapp. P.-v. Réun. Cons. int. Explor. Mer 183:255-268.

Powers, K. D. In press. Estimates of consumption by seabirds on Georges Bank. In Georges Bank. Ed. by R. H. Backus. MIT Press, Cambridge.

Scott, G. P., Kenney, R. D., Thompson, T. J., and Winn, H. E. 1983. Functional roles and ecological impacts of the cetacean community in the waters of the northeastern U.S. continental shelf. ICES C.M.1983/N:12.

Sherman, K., Jones, C., Sullivan, L., Smith, W., Berrien, P., and Ejsymont, L. 1981. Congruent shifts in sand eel abundance in western and eastern North Atlantic ecosystems. Nature 291:486-489.

Sherman, K., Smith, W. G., Green, J. R., Cohen, E., Berman, M. S., Marti, K., and Goulet, J. R. In press. The invertebrate zooplankton and ichthyoplankton. In Georges Bank. Ed. by R. H. Backus. MIT Press, Cambridge.

Sissenwine, M. P. In press. Fish production. In Georges Bank. Ed. by R. H. Backus. MIT Press, Cambridge.

Sissenwine, M. P. 1984. Why do fish populations vary? In Exploitation of marine communities. pp. 59-94. Ed. by R. M. May. Dahlen Konferenzen. Springer-Verlag, Berlin.

Sissenwine, M. P., Cohen, E. B., and Grosslein, M. D. 1984. Structure of the Georges Bank ecosystem. Rapp. P.-v. Réun. Cons. int. Explor. Mer 183:243-254.

Slobodkin, L. B. 1961. Growth and regulation of animal populations. Holt, Reinhart, and Winston, N.Y. 184 pp.

Part 2

Measuring Variability
in Large Marine Ecosystems

6. Definitions of Environmental Variability Affecting Biological Processes in Large Marine Ecosystems

ABSTRACT

Attempts to relate variations in reproductive success of fish populations to variability in the environment or in the biological community have been generally unsuccessful. A variety of processes, varying widely over a broad range of time and space scales, is potentially involved. Continuous observational coverage over such a range is impossible. Under the circumstances, defining the crucial variations involved presents a challenging set of problems. A serious need exists to develop generalizations that might provide the basis for unifying the available fragments of experience in a usefully coherent framework so as to make optimal use of the few empirical data available in a given situation. Interregional comparative studies, systematic observations of vertical distributions of organisms, studies of within-year variability in survival, and use of time/space statistical techniques, are advocated.

INTRODUCTION

In terrestrial ecosystems, the atmosphere has little capacity to store heat and other properties, and most sources of materials, etc., are fixed to a nonmoving substrate. In contrast, marine ecosystems exist in a continually moving, nonhomogeneous, dense fluid which has a large capacity to store heat, materials, and momentum. As a result, marine ecosystems are characterized by long period, large amplitude variability. During recent decades, catastrophic declines of some of the world's largest fishery stocks have had severe human consequences. During 1983 the ecosystems of the eastern Pacific were experiencing the most intense environmental anomalies of the past half century; it is likely that transient responses associated with this "El Niño" event will continue to perturb these ecosystems for years to come. The present lack of understanding of the mechanisms governing the interaction of environmental variation and fishery exploitation in causing large fluctuations in "recruitment" (i.e., the numbers of younger fish entering a fished population) is recognized as perhaps the most serious scientific problem hindering more effective management of living marine resources. Discussion in this chapter focuses on examples from eastern boundary current pelagic fishery stocks, as being illustrative of research problems common to a variety of large marine ecosystems and associated biological communities.

THE DILEMMA

Demonstrating that change occurs in these environmental/biological systems is not difficult. But to qualitatively and quantitatively specify the changes in a way that can form a basis for empirical understanding of the linkages involved presents a challenging suite of problems. Since the number of fish surviving to adulthood may be four orders of magnitude lower than the initial number of newly hatched larvae, such eventual survival is quite a rare event for individuals. Thus spot sampling of small numbers of realizations of individual life cycle events would be unlikely to yield useful information about total recruitment of a particular stock unit, even if the stock were homogeneously distributed and survival processes were linear. In fact, nonlinear interactions which are highly variable over a broad range of time and space scales typify the growth and mortality processes of pre-recruit fishes and most other marine organisms. Thus the ability to identify particularly crucial environmental conditions and to properly integrate their effects over the reproductive region and season associated with a stock unit, so as to reflect the net effect on reproductive success, appears to be a prerequisite for effective empirical approaches to understanding biological variability in large marine ecosystems.

Crucial Time Scales

Certain time scales appear to be of particular interest. For example, Lasker (1978) has indicated that anchovy larval survival may be seriously impaired by dispersion of fine-scale strata of microorganisms, required for food of "first-feeding" larvae, by wind-related turbulent mixing. Assuming that the time for accumulation of such concentrated strata is longer than one day, the relevant temporal considerations involve the interaction of the several days required for larval starvation, atmospheric storm activity with periods of variation of the order of a day to several weeks (examples shown in Figure 6.1), and the temporal variation of spawning activity over the spawning season.

At somewhat longer time scales (several weeks to several months) a variety of mechanisms may have significant effects. These include advection to favorable or unfavorable habitat, match/mismatch of spawning peak with food or predator organisms, interaction of temperature or food-regulated growth rates with size-dependent predation, adult feeding conditions which may affect formation of reproductive products.

The annual periodicity tends to be very dominant (Figure 6.1c); even in equatorial regions seasonal monsoon effects may be important. Biological processes tend to be well tuned to the annually repeating environmental variations, such that anomalies from normal climatic conditions are likely causes of disruptions in reproductive success of endemic populations. The dominance of the characteristic annual cycle of variation (compared to anomalies from it) causes problems in empirical analysis. For one thing, the degrees of freedom in within-year series available for empirical analysis tend to be substantially reduced because of the difficulties involved in properly filtering the large annual signal. Perhaps even more crucial is the fact that responses by natural selection to the large seasonal variation have probably, in many cases, resulted in seasonal tuning of basic biological--environmental linkages, such that underlying relationships among variables may be highly non-stationary over the annual cycle. These difficulties have motivated the conventional practice of studying recruitment processes on the basis of annual composites of the shorter scale events which may be actually involved.

On interyear time scales, marine biological time series very often exhibit substantial variance in the 3- to 12-yr period band (e.g., Figure 6.2). This is a band corresponding to appropriate periods of variation for various oceanic and atmospheric phenomena including baroclinic Rossby waves (Mysak et al., 1982) and variations of the global coupled ocean/atmosphere system known as El Niño-Southern Oscillation (ENSO) events. Other classes of variations also occupy this band (e.g., sunspots, etc.) and are sometimes incorporated into empirical "models" of biological variation; this is a band where the available time series is generally too short to contain more than a very few realizations of such multiyear variations and particularly in cases where direct linkage mechanisms are not well defined, the likelihood of spurious relationships is high.

92

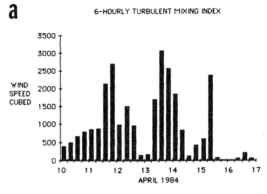

a

6-HOURLY TURBULENT MIXING INDEX

WIND SPEED CUBED

APRIL 1984

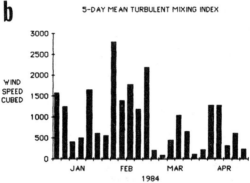

b

5-DAY MEAN TURBULENT MIXING INDEX

WIND SPEED CUBED

JAN   FEB   MAR   APR
1984

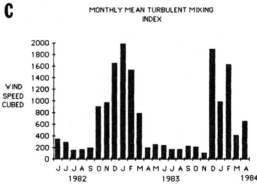

c

MONTHLY MEAN TURBULENT MIXING INDEX

WIND SPEED CUBED

J J J A S O N D J F M A M J J A S O N D J F M A
1982   1983   1984

Figure 6.1. Index of rate of addition of turbulent mixing energy to the water column by the wind (e.g., Husby and Nelson, 1982). The index is computed as the third power (cube) of the speed of the large scale wind, as derived from synoptic surface pressure/wind analyses obtained from Fleet Numerical Oceanography Center. The location illustrated is off the coast of Washington State, at 48°N 125°W. Units are m^3sec^{-3}. (a) 6-hourly synoptic computations. (b) 5-day means of 6-hourly computations. (c) Monthly means.

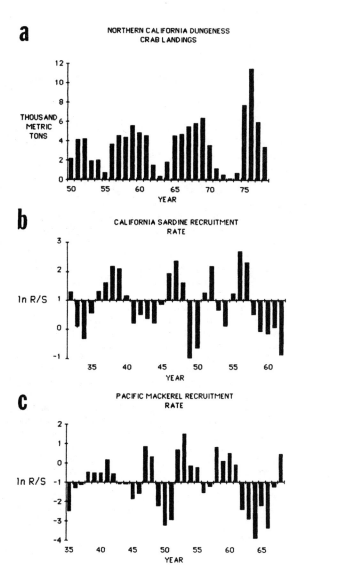

Figure 6.2. (a) Commercial fishery landings of dungeness crab in northern California (after Wild,
1983). Units are thousands of metric tons. Year
indicated refers to the calendar year containing the
end of a given annual fishing season. (b) Natural log
of the ratio of California sardine recruits (10^6 fish)
to parental biomass (10^3 metric tons). Values for
1947-1962 are based on the estimates of MacCall (1979).
Values for 1932-1946 are based on the estimates of Murphy (1966). (c) Natural log of the ratio of Pacific
mackerel recruit biomass to parental biomass (after
Parrish and MacCall, 1978).

It is becoming apparent that there may be hierarchies of El Niño episodes, such that the events of 1957-1958 and 1982-1983 stand out as being global in character. These two episodes represent such extreme outliers in many California Current biological time series that they tend to control relationships based on "least square" fitting procedures. There is reason to believe that basic mechanisms underlying relationships among variables may be altered during these extreme episodes. Therefore, analysis of data from these particular years, together with data from the more normal years, may obscure any real relationships existing in either subset. The period between these two extreme episodes is 25 yr; this period is comparable to the total length of most pertinent data series from the California Current ecosystem. Deriving any substantial insight from an empirical analysis based on as little as two actual realizations is difficult, to say the least.

An even longer time scale may be highly pertinent to fishery management considerations. Smith (1978) has pointed out indications that the total biomass of pelagic fish stocks in the California Current system may have been as much as several times larger in the early part of this century than at the present time, or even than in the late 1930s when heavy exploitation commenced (Figure 6.3). This poses some important questions. Does the presence of large fish populations somehow increase the ability of the ecosystem to support the maintenance of these large populations? Do massive shifts in partitioning of basic organic production among exploitable and non-exploitable ecosystem components occur? Can management actions influence these shifts? Obviously, this is a time scale that we are not prepared to address experimentally and must rely on available "proxy" records, such as fish scale deposition rates (Soutar and Issacs, 1974) or guano deposits (e.g., Crawford et al., 1983), for empirical data analysis.

Crucial Space Scales

Lasker's hypothesis involves fine-scale vertical strata which may be centimeters or less in thickness. Crucial horizontal scales include spawning patch, food organism patch, juvenile school, small eddy, surface convergence zone, etc., diameters which may range from meters to kilometers (Smith, 1978; Owen, 1981). Offshore transport of reproductive products is thought to be an important regulator of reproductive success in eastern boundary systems (Parrish et al., 1983); pertinent offshore length scales are thought to include the internal Rossby radius of deformation (10-20 km off

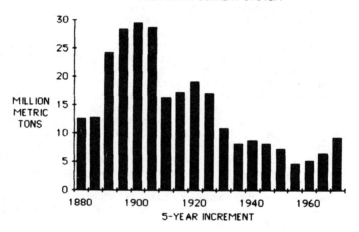

TOTAL COASTAL PELAGIC FISH BIOMASS IN THE
CALIFORNIA CURRENT SYSTEM

Figure 6.3. Biomass of the total coastal pelagic fish
complex, consisting of sardine, anchovy, mackerel,
saury and hake, in the California Current system at 5-
yr increments, as inferred by Smith (1978) from analysis of present day stock sizes and from scale deposition rates determined by Soutar and Issacs (1974).

northern California, about four times greater in the
anchoveta spawning region off northern Peru) and the
width of the continental shelf (e.g., 20 km off n.
California, 130 km off n. Peru). Major coastline
features may delineate upwelling centers and coastal
gyral circulations (length scales of tens to hundreds
of kilometers). Spawning regions of coastal pelagic
stocks may be of the order of a thousand kilometers in
coast wise extent and several hundred kilometers in
offshore extent. Larval drift and adult migrations may
cover a thousand or more kilometers.

Environmental Data

    Continuous observational coverage of such a range
of time and space scales, using dedicated research
ships, is not possible. One hope is to make use of
routinely-gathered maritime weather reports which are
available at 6-hourly synoptic intervals according to
international convention. Historical files extend into
the past century. Thus coverage is available over the
range of desired time scales. However spatial resolution leaves much to be desired (Figure 6.4). Of some
advantage is the fact that atmospheric length scales
are large compared to oceanic length scales, and thus
the major spatial scales of variation of wind, cloud
cover (i.e., solar radiation input), etc., are at least
minimally covered. Sea surface temperature varies
relatively slowly compared to atmospheric properties
and therefore observations from a number of synoptic
samplings may be composited to improve coverage of
finer-scale spatial features.
    Fortunately, several currently-favored hypotheses
concerning environmental regulation of reproductive
success feature mechanisms involving wind, sea temperature, and solar radiation. In Lasker's (1978)
hypothesis, wind-generated turbulent mixing energy is
the causative factor. Wind-driven offshore surface
transport, causing offshore loss of reproductive
products (Parrish et al., 1983) is related to the
product of the wind speed with its alongshore velocity
component. On the other hand locally wind-induced
coastal upwelling is driven by this same offshore
surface transport which, therefore, may favor reproductive success at appropriate temporal and spatial
lags by enhancing primary production and leading to
adequate larval food particle concentrations, while
being an unfavorable factor when coincident with
spawning activity. Solar radiation also is involved in
primary production, while sea temperature has direct
effects on physiological rates; since larvae of many
species inhabit the surface mixed layer, surface
temperature often provides a useful indicator of

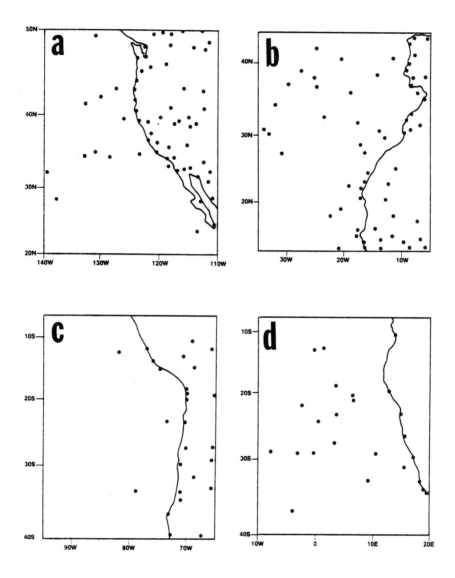

Figure 6.4. Typical synoptic report distributions in the four major eastern ocean boundary upwelling regions (after Bakun and Parrish, 1980). (a) The California Current region (distribution shown is for 0600 GMT, 11 Jan 1980). (b) The Canary Current region (0000 GMT, 12 Feb 1980). (c) The Peru Current region (1800 GMT, 7 Feb 1980). (d) The Benguela Current region (1800 GMT, 7 Feb 1980).

temperature effects. Wind stress curl drives upwelling and downwelling seaward of the coastal boundary zone, controls convergence and divergence in the surface Ekman layer, and through Ekman pumping is a primary driving agent for variations in the baroclinic structure of the upper ocean (which may subsequently propagate westward as Rossby waves of interannual time scale).

Precision of estimates based on summaries of scattered maritime reports is a matter of concern. Since our focus is generally on interyear variability, difficulty in perfectly filtering the strong annual cycle exacerbates the low signal-to-noise ratio problem. Comparisons of random subsamples from a ten-degree latitude/longitude quandrangle adjacent to a typical temperate eastern boundary coast (Bakun, in preparation) indicate that only the more extreme interyear variations may be definable by simple areal summaries of reports at typical densities available. For example, it appears that at 200 observations per monthly sample we can expect that only about 20% of interyear differences (i.e., the difference between a monthly value and the value for the same month one calendar year earlier) of wind-mixing index (3rd power of wind speed) will be significant at the 0.1 level. For summaries of alongshore wind stress and cloud cover the situation appears to be somewhat better; about 60% of interyear differences in these properties appear to be significant at this level at a similar data density.

Additional information can be incorporated by means of objective analysis techniques (e.g., Holl and Mendenhall, 1972). These techniques spread information in time and space in a physically realistic manner, using dynamic or statistical models. Thus, observations from previous and subsequent synoptic samplings or from neighboring locations may contribute to a better estimate of the expected value at a gap in data coverage than possible from simple interpolation. Objective data analyses produced routinely by national meteorological agencies generally also incorporate elaborate error checking and quality control procedures based on degree of conformity of observations to physical reality as defined by the models. In addition large scale analysis of wind observations may be improved by the incorporation of barometric pressure data through the geostrophic coupling which is quite strong on large scales in extratropical regions. Thus, the physical mechanisms underlying the models, and the temporally and spatially adjacent data, constitute additional information which may improve the estimates. Thus, we may expect that significance of indicated interyear differences in monthly values computed from objectively analyzed fields (e.g., wind mixing index as shown in Figure 6.1c) may be improved relative to simple areal averages of reports. However, the complexity of the analysis procedures make it difficult to quantify the expected degree of improvement.

Monthly "coastal upwelling indices" based on analyses of wind and atmospheric pressure, are intended to indicate interyear variations in wind-induced offshore transport of surface waters and resulting coastal upwelling. They have been found to yield significant correlations with a wide range of types of biological variations in the California Current region (Bakun and Parrish, 1980), involving such fishery species as Pacific sardine, northern anchovy, Pacific mackerel, dungeness crab, English sole, coho salmon, bonito, hake, and shrimp. However, precision estimates for the indicated variations are not available and associated difficulties in assigning confidence estimates on the apparent relationships limit their utility.

Specifying shorter-scale variations, in the energetic "storm event" frequency band (e.g., Figures 6.1b and 6.1c), is less of a problem due to the large amplitude of the signal relative to the noise (almost any sort of analysis of reports is adequate to indicate presence or absence of storm conditions). However, specification of the affected biological characteristics is seldom available on corresponding scales, and so the tendency has been to search for "bulk" relationships which integrate larger (but unfortunately lower in amplitude) scales of variation.

Measurements at coastal installations offer useful sources of data, covering desirable ranges of time scales at consistent geographic locations. Spatial aspects are less satisfactory. Coastal sea temperature measurements may sample very local scales of variation. Wind measurements are often distorted by coastal topography. Sea level measurements do reflect rather large scales but these scales are difficult to specify, which is a large problem because the main effect of sea level on the processes of interest is through its gradient.

Statistical Significance

In summary, we face a problem of substantial complexity with very little empirical information covering the range of time and space scales determining the final level of recruitment to a total stock. As long as we must attempt to relate "bulk" effects on larger scales (with lower signal amplitudes) than those on which the crucial processes are acting, we must expect weak relationships which do not have a high degree of statistical significance given the limited degrees of freedom available on interyear scales in fishery and other biological time series. In most types of situations it would be almost inconceivable that any one environmental mechanism could so consistently dominate survival processes as to yield a strong univariate relationship that was not, to some degree, spurious. This is not to say that a well-formulated exploratory data analysis is not a useful activity, or that its results should not be reported. But mere searches for relationships meeting significance criteria (i.e., publishable), among various combinations of data sets and formulations, are not very enlightening. The question of how to make use of the weak relationships, which may represent realistic levels of control by particular mechanistic linkages, poses a major challenge.

SOLUTIONS

An obvious need is to develop generalizations that might apply to wider classes of fisheries or ecosystems, and thereby provide a basis for unifying the available fragments of experience in a usefully coherent framework. Certainly, if each separate situation is fundamentally different from all others, there does not seem to be much hope of developing substantial understanding of any. For that matter, there would probably be few situations worth intense research effort if expected results were to have no broader sphere of application. The following paragraphs present some suggestions as to approaches which might lead to progress in resolving some of the problems outlined above.

Interregional Comparative Studies

The large marine ecosystems of the four major subtropical eastern ocean boundary regions appear to be controlled by similar environmental dynamics and contain very similar assemblages of exploitable pelagic fishes (Table 6.1). They also exhibit similar histories of fishery growth and abrupt decline (Figure 6.5). The similarities suggest that the fish communities in these different systems may function somewhat similarly with respect to their environments and may have reproductive strategies adapted to solving similar environmental problems. Since natural selection implies that reproductive strategies reflect responses to the most crucial factors regulating reproductive success, compelling patterns of correspondence among reproductive habits and environmental characteristics are likely to reflect important causal mechanisms.

For example, Parrish et al., (1983) examined characteristic seasonal distributions of temperature,

TABLE 6.1.

Dominant anchovy, pilchard, horse mackerel, hake, mackerel, and bonito in the four major eastern boundary currents (after Bakun and Parrish, 1980).

California Current	Peru Current	Canary Current	Benguela Current
Engraulis mordax	Engraulis ringens	Engraulis encrasicholus	Engraulis capensis
Sardinops sagax	Sardinops sagax	Sardinia pilchardus	Sardinops ocellatus
Trachurus symmetricus	Trachurus symmetricus	Trachurus trachurus	Trachurus trachurus
Merluccius productus	Merluccius gayi	Merluccius merluccius	Merluccius capensis
Scomber japonicus	Scomber japonicus	Scomber japonicus	Scomber japonicus
Sarda chiliensis	Sarda chiliensis	Sarda sarda	Sarda sarda

102

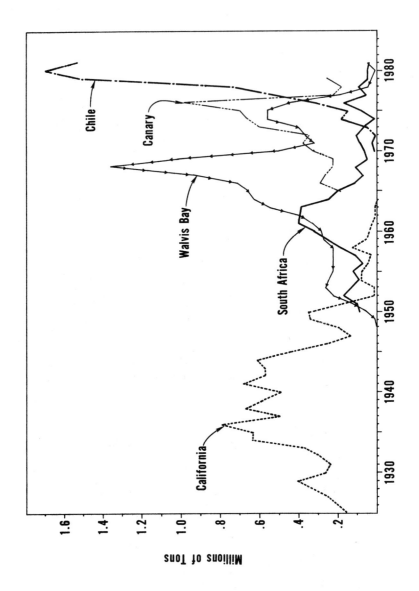

Figure 6.5. Annual sardine landings from several
eastern boundary current stocks (after Parrish et al.,
1983). Walvis Bay and South Africa refer to the two
major Benguela Current stocks.

surface drift, wind mixing energy input, and upper
water column stability in the California, Peru, Canary,
and Benguela Current systems. They found several
interesting patterns with respect to available information on spawning of anchovies and sardines. Avoidance
both of substantial wind mixing of the upper ocean
layer and of strong offshore transport seemed to be
dominant characteristics of spawning habitat selection.
Selection for any particular optimum temperature, within physiological limits, seemed less important. These
findings, which are independent of time-series results,
provide a guide for selection of variables for time-
series modelling in a way that makes improved use of
the scarce degrees of freedom available.

Certain initially-puzzling discrepancies from
general patterns have proven enlightening. For
example, it was noted that the spawning peak of the
Peruvian anchoveta occurred in austral winter, the
season of strongest offshore surface transport, rather
than at the season of weakest transport, as is the
general case. However, when viewed in conjunction with
mixed layer depth climatology the discrepancy appears
to be resolved; seasonal variation of mixed layer depth
off Peru proceeds in phase with that of transport but
has greater amplitude, with the result that the thinner
mixed layer of austral summer is apparently carried
offshore nearly four times as fast as is the deeper
winter mixed layer, even though the winter volume of
transport is nearly twice as large. Thus spawning
indeed appears tuned to minimize the rapidity of offshore loss of eggs and larvae. This resolution of an
initial discrepancy points out that, in treating the
mechanism of offshore loss of reproductive products
which are distributed through the mixed layer, wind-
driven surface (Ekman) transport should ideally be
divided by mixed layer depth to yield an Ekman velocity
of the mixed layer. In this way, two environmental
variables are combined to yield a single more
meaningful variable, thereby preserving degrees of
freedom for fitting additional pertinent variables in
empirical modelling efforts.

Beyond comparisons of spawning habitat climatology, it would seem to be fruitful to compare actual
time-series relationships among similar regional
ecosystems. Under an assumption of analogy, weak
empirical relationships having a similar general form
in several systems could be assigned greater confidence
than would otherwise be warranted. Conversely, a high
correlation found in one system might be questioned if
no suggestion of a similar type of linkage was found in
other systems appearing in other respects to be analogous. Arraying identically formulated empirical models
from different regional systems might yield patterns

among model parameters that could provide new insights as to proper transfer of experience among systems in order to predict outcomes of actions taken in one system from the record of similar actions taken previously in other systems. Such efforts would ideally involve interregional and international cooperative activities to assemble analogous variable sets, and provide informed interpretation.

Observational Activities Aiding Scale Integration

Comparative studies are an example of a deductive (i.e., indirect) approach. Because of the great range of crucial scales, more direct experimental approaches tend to involve an impractical observational investment. However, it may be possible to save some effort by designing sampling schemes that would enhance our interpretation of larger scale, more easily specified variations in terms of their effects on processes occurring at smaller scales. As an example, it is conceivable that weakly swimming organisms might exert considerable control over their advection by simply adjusting their depth in the water column. In subtropical eastern ocean boundary regions, clear weather often accompanies an offshore atmospheric high pressure system which produces equatorward winds and corresponding offshore-directed surface Ekman transport. Stormy weather would involve the opposite conditions leading to onshore surface transport. Thus, organisms that position themselves deeper in the water column during clear weather than under cloudy conditions would be aided in avoiding detrimental offshore advection. In addition, the very surface "skin" of the ocean travels in a different direction with respect to the wind than does the Ekman layer; other surface skin effects may act somewhat independently of the wind, e.g., onshore movement of surface slicks generated by passage of internal waves (Shanks, 1983). Knowledge that has been developed concerning the physical transport mechanisms would allow interpretation of vertical biological structure in terms of the function and importance of these mechanisms in natural selection. It appears that well-determined characteristic vertical distributions of planktonic organisms are not widely available, although the sampling techniques are not extraordinarily difficult (e.g., compared to quantitative population estimation). Some detailed sampling of the vertical distributions under different conditions, which might constitute a practical degree of effort, could reveal much about the manner of biological response to these conditions. With increased understanding of these responses, monitoring of the larger

scale conditions might in some way substitute for
continuous time/space observation of the actual fluxes
of organisms.

Sampling of "Integrated" Effects
at "High Signal" Scales

The logical way to seek the greatest empirical
return for an observational investment is to sample the
integrated results of processes occurring on smaller
scales, which are more readily sampled but where the
signal-to-noise ratio remains relatively high. Conventional recruitment estimates derived from fishery
statistics integrate within-year variations to yield a
single annual composite. This process suppresses and
obscures signals from shorter scale, possibly higher
amplitude, components of the variability in survival.
In addition, the process of forming composites necessi-
tates an integration of energetic environmental "event"
frequencies with accompanying decline in signal-to-
noise ratio in specifying the environmental context.
Working with annual composites also yields minimal
degrees of freedom for empirical investigation of
linkages.

In connection with the International Recruitment
Project (IREP) of the Intergovernmental Oceanographic
Commission's Program of Ocean Science in relation to
Living Resources (OSLR), an experimental program has
been conceived (Anon., 1983) which appears to be
uniquely promising because it addresses the energetic
within-year scales. It involves determining fish egg
or larval abundance within a spawning region of a given
stock unit, at increments through an extended spawning
season. Subsequent survival is estimated by sampling
late larvae or early juveniles, determining their
birthdate distributions using recently-developed daily
ageing techniques (Methot and Kramer, 1979), and comparing birthdate distributions to the corresponding
estimates of egg abundance. Interference from
annually-tuned mechanisms is minimized by the process
of incrementally-determining both the inputs (eggs or
early larvae) and the outputs (survival). Integrations
of environmental conditions over the intervening time
increments can then be examined to determine the degree
of "fit" represented by alternate causal mechanisms and
formulations. The program does imply a considerable
sampling effort to adequately encompass the pertinent
scales of variation over a total stock unit.

Time/Space Statistical Analysis

With the rapid increase in economical computing
power, application of a very powerful class of methods

of exploratory data analysis is finding increased application in fishery/environment studies. Basically, they allow the computer to do much of the work in sorting out time/space structures among data sets (work that may previously have been done less systematically by laboriously plotting and comparing time series of mapped distributions). For example, Mendelssohn and Roy (in press) have used spectral empirical orthogonal functions to demonstrate wavelike propagation in tuna distributions and in sea temperature which conforms to a Kelvin wave interpretation. Such examinations of space/time structures in the behavior of ecosystem components, in the light of known structures in the characteristics of particular environmental mechanisms, appear to offer substantial promise. However, it may be well to note that rather than being a mechanical process, both the formulation of the variables and the application of the methods involve important interpretive aspects requiring skill and judgment; without these, the increasing ease of access to electronic data banks, substantial computing power, and software for complex analysis procedures, would appear to have potential for generation of a much greater volume of published output, without any correspondingly greater degree of enlightenment, than the correlation searches of the past.

REFERENCES

Anon. 1983. Summary report. Workshop on the IREP Component of the IOC Program on Ocean Science in Relation to Living Resources (OSLR). Intergovernmental Oceanographic Commission, UNESCO Paris. IOC Workshop Rep. 33:1-17.

Bakun, A., and Parrish, R. H. 1980. Environmental inputs to fishery population models for eastern boundary current regions. In Workshop on the effects of environmental variation on the survival of larval pelagic fishes. Ed. by G. D. Sharp. Intergovernmental Oceanographic Commission, UNESCO, Paris. IOC Workshop Rep. 28:67-104.

Crawford, R. J. M., Shelton, P. A., and Hutchings, L. 1983. Aspects of variability of some neritic stocks in the southern Benguela system. In Proceedings of the expert consultation to examine changes in abundance and species of neritic fish resources. Ed. by G. D. Sharp and J. Csirke. FAO Fish. Rep. 291(2):407-448.

Holl, M. M., and Mendenhall, B. R. 1972. Fields by information blending, sea level pressure version. Fleet Numerical Oceanography Center, Monterey, Calif. FNWC Tech Note 72-2. 66 pp.

Husby, D. M., and Nelson, C. S. 1982. Turbulence and vertical stability in the California Current. Calif. Coop. Oceanic Fish. Invest. Rep. 23:113-129.

Lasker, R., 1978. The relation between oceanographic conditions and larval anchovy food in the California Current: Identification of factors contributing to recruitment failure. Rapp. P.-v. Réun. Cons. int. Explor. Mer 173:212-230.

MacCall, A. D. 1979. Population estimates for the waning years of the Pacific sardine fishery. Calif. Coop. Oceanic. Fish. Invest. Rep. 20:72-82.

Mendelssohn, R., and Roy, Cl. (In press). Environmental influences on the F.I.S.M tuna catches, in the Gulf of Guinea. Special Volume of the International Skipjack Year Program. International Commission for the Conservation of Atlantic Tuna (ICCAT), Madrid.

Methot, R. D., and Kramer, D. 1979. Growth of northern anchovy (Engraulis mordax) larvae in the sea. Fish. Bull., U.S. 77:413-423.

Murphy, G. I. 1966. Population biology of the Pacific sardine (Sardinops caerulea). Proc. Calif. Acad. Sci. 34:1-84.

Mysak, L. A., W. W., Hsiah, and Parsons, T. R. 1982. On the relationship between interannaul baroclinic waves and fish populations in the northeast Pacific. Biol. Oceanogr. J. 2:63-103.

Owen, R. W. 1981. Microscale plankton patchiness in the larval anchovy environment. Rapp. P.-v., Réun. Cons. int. Explor. Mer 178:364-368.

Parrish, R. H., Bakun, A., Husby, D. M., and Nelson, C. S. 1983. Comparative climatology of selected environmental processes in relation to eastern boundary current pelagic fish reproduction. In Proceedings of the expert consultation to examine changes in abundance and species of neritic fish resources. Ed. by G. D. Sharp and J. Csirke. FAO Fish. Rep. 291(3):731-778.

Parrish, R. H., and MacCall, A. D. 1978. Climatic variation and exploitation in the Pacific mackerel fishery. Calif. Dep. Fish Game, Fish Bull. 167:109.

Shanks, A. L. 1983. A mechanism for the onshore migration of pelagic larvae of invertebrates and fish: Transport in slicks over tidally induced internal waves. (Abstract 42B-01). EOS-Trans. Am. Geophys. Union 64:1080.

Smith, P. E. 1978. Biological effects of ocean variability; Time and space scales of biological response. Rapp. P.-v. Réun. Cons. int. Explor. Mer 173:117-127.

Soutar, A., and Issacs, J. D. 1974. Abundance of pelagic fish during the 19th and 20th centuries as recorded in anaerobic sediment off the Californias. Fish Bull., U.S. 72:257-273.

Wild, P. W., Law, P. M. W., and McLain, D. R. 1983. Variations in ocean climate and the Dungeness crab fishery off California. In Life history, environment, and mariculture studies of the dungeness crab, Cancer magister, with emphasis on the Central California fishery resource. Ed. by P. W. Wild and R. N. Tasto, Calif. Dep. of Fish Game, Fish Bull. 172:175-188.

7. Variability of the Environment and Selected Fisheries Resources of the Eastern Bering Sea Ecosystem[1]

ABSTRACT

The commercial harvest of fish and invertebrates from U.S. waters of the eastern Bering Sea shelf ecosystem comes primarily from south of 60°N, with an average yield from this area exceeding 4 metric tons km^{-2} yr^{-1}. More than 70% of the total harvested biomass is comprised of a single species, walleye pollock (Theragra chalcogramma). Three species of benthic crab account for a small proportion of total harvest by weight, but are economically important and show marked fluctuations in abundance over periods of a few years. Spatial and temporal patterns of abundance of these four species are examined, emphasizing the occurrence of particularly strong year classes. Inter-annual variability in recruitment of these species is discussed with respect to prominent, variable features of their biotic and abiotic environment. Contrast is provided by two congeneric species of crab exhibiting markedly different recruitment patterns. The principal physical variables are sea temperature, horizontal, and vertical mixing (particularly during spring), and mean transport over periods of one or more months. The biotic variables for which we have data related to these species are zooplankton abundance during spring and changes in abundance of major predators in the ecosystem. The number of years of observation is insufficient to assess the real importance of any of the proposed mechanisms; however, the environmental and organismal changes are marked and well suited to further study.

INTRODUCTION

The expansive shelf of the eastern Bering Sea (Figure 7.1) harbors a rich assemblage of fish and invertebrates which are permanent members of the shelf ecosystem, though they may migrate seasonally to deeper

110

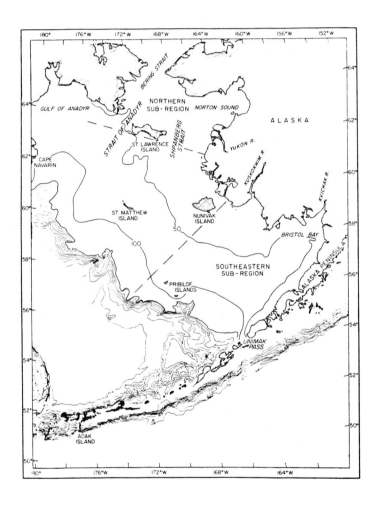

Figure 7.1. Geographic names and prominent bathymetric contours of the eastern Bering Sea shelf (depths in meters).

or shallower regions of the shelf for overwintering or spawning. Of the species which are commercially harvested, most are demersal, semidemersal, or benthic. Only one is truly pelagic, Pacific herring (Clupea harengus pallasi), and this species generally accounts for less than 2% of the total harvested biomass. In addition to the resident species, marked seasonal migrations of salmon into and out of the ecosystem occur (Straty, 1981). These migrations are ecologically significant and are the basis of a large commercial fishery; however, salmon are not year round residents of the shelf and will not be considered further in this chapter.

The major resident species of fish and invertebrates harvested from the eastern Bering Sea are listed in Table 7.1, along with commercial landings reported for 1981. Nearly all harvests were from shelf and slope areas south of 60°N, and thus from an area of approximately 311,245 km^2. The total harvest from 1981 (including only that portion of the herring stocks from south of 60°N and excluding the harvest of salmon) equates to a mean commercial yield of 4,298 kg or 4.7 tons km^{-2} for the year. This yield represents only a portion of production at upper trophic levels. Species such as capelin (Mallotus villosus) and sand lance (Ammodytes hexapterus) may play a significant role in the ecosystem's energy, but these species are not fished and remain poorly sampled. While restricted in scope, commercially harvested species provide most of the data by which we gauge variability in production at upper trophic levels.

Fisheries of the eastern Bering Sea shelf are managed with the goal of maintaining, as much as possible, stable yields of total harvest from the multispecies complex. This approach shifts harvest allocations among species depending on the prevailing condition of various stocks. Knowledge of how species and stocks interact, and how the abundance of a species is affected by other biotic and abiotic factors is thus of practical value in management just as it is of scientific interest. It seems fundamental that we ultimately should be able to distinguish between fluctuations in abundance arising from conditions which we can modify (for example, through redirected fishing efforts), and those resulting from events which cannot be managed (such as variations in climate and production at lower trophic levels).

In this chapter, we focus on stocks of four commercially harvested species from the eastern Bering Sea, examine changes in abundance with time, and consider possible key interactions between these stocks and their biotic and abiotic environment. Since fluctuations in biomass frequently are manifestations

TABLE 7.1.

Reported major commercial harvests of resident shelf species from the eastern
Bering Sea during 1981. Harvests are primarily from the shelf and south of
60°N except as otherwise noted (Sources: National Marine Fisheries Service
and Alaska Department of Fish and Game).

Species		Harvest
Common Name	Generic Name	(thousands of kg live weight)

I. Demersal and semi-demersal species:

Walleye pollock	Theragra chalcogramma	973,505
Yellowfin sole	Limanda aspera	97,301
Pacific cod	Gadus macrocephalus	59,916
Greenland turbo[a]	Atherestes stomias	52,921
Arrowtooth flounder[a]	Rheinhardtius hippoglossoides	13,473
Rock sole	Lepidopsetta bilineata	~9,021[b]
Alaska plaice	Pleuronectes quadrituberculatus	~8,653[b]
Flathead sole	Hippoglossoides elassadon	~5,193[b]
Squid	(mostly Berryteuthis magister and Onchoteuthis borealijaponicus)	~4,159
Atka mackerel	Pleurogrammus monopterygius	3,028
Sablefish[a]	Anoplopoma fimbria	2,600
Rockfishes	Sebastes spp. and Sebastolobus spp.[c]	2,402
Misc. flatfishes	(several genera; see Bakkala, 1983)	~ 561

II. Pelagic species:[d]

Pacific herring	Clupea harengus pallasi	18,291[e]

III. Miscellaneous fish and cephalopods:[f]   35,551

IV. Benthic crabs:

Tanner crabs	Chionoecetes bairdi and C. opilio[g]	53,002
King crabs	Paralithodes camtschatica and P. platypus	3,340

Total fish and molluscs	1,286,575
Total crab	56,342
Total commercial harvest	1,342,917

[a]These species are mostly from the slope.

[b]A small proportion of these reported harvests were from the Aleutian
Islands area.

[c]Approximately 52% of this was Pacific Ocean Perch, Sebastes aluta.

[d]Salmon harvests are not included in this list because the adult fish
are only temporarily associated with the shelf. Landings from the eastern
Bering Sea south of 60°N during 1981 were 93,730,000 kg.

[e]Of this total, approximately one-fifth (3,964,000 kg) came from north
of 60°N. Herring are presently harvested from very shallow (generally less
than 20 m) water during their spawning migration, but spend most of the year
over deeper areas of the shelf (Wespestad, 1981).

[f]Species presently of low economic values, generally not targeted:
sculpins, skates, smelts and octopuses.

[g]Includes a small number of hybrid (C. bairdi x C. opilio) Tanner crabs
(see Otto, 1981; Johnson, 1976).

of year-class abundance, we concentrate on factors which have impacts on this time scale. We stress those aspects of recruitment which might show direct or indirect responses to physical environmental changes. Descriptions of the physical environment and the four stocks precede our discussion. Spatial and temporal (seasonal and interannual) variability in production and/or standing stocks of phytoplankton and zooplankton have been reported by Iverson et al. (1979), Cooney and Coyle (1982), Walsh (1983), Sambrotto and Goering (1983, 1984), Smith and Vidal (1984, in press), and Vidal and Smith (in press).

PHYSICAL ENVIRONMENT

The eastern Bering Sea shelf has two unique physical features: its vast size and a seasonal ice cover. A "leak" at its northern boundary, the Bering Strait, has significant influence on flow over the shelf. This shelf, the largest of the World Ocean outside the Arctic, is bounded on the south by the Alaska Peninsula and on the north by Alaska and Siberia. It exceeds 500 km in width at its narrowest point and subtends 11° of latitude. The shelf deepens very gradually (over three-fourths is contained within the 100-m isobath) to about 170 m at the shelf break. Here the bathymetry is complex and indented by several large canyons. Although the shelf is generally featureless, enhanced bottom slope in the vicinity of the 50- and 100-m isobaths affects currents and property distributions. Ice cover is seasonal, varying from none between late June and November to greater than 80% coverage of 0.5 to 2.0 m thick ice during its maximum extent in March (Niebauer, 1980; Pease, 1980). Ice is all first-year ice, primarily formed along leeward (south-facing) coasts and advected by winds to its thermodynamic limit over the outer shelf/slope. Interannual variations in extent can be hundreds of kilometers (Overland and Pease, 1982), or more than 60% of the mean coverage (Niebauer, 1983). Averaged over a year, approximately one Sverdrup (10^6 m^3/s) flows northward through Bering Strait (Coachman and Aagaard, 1981). Some of this transport (less than one-fourth) is accounted for by inflow of coastal water (the Kenai Current) from the Gulf of Alaska through Unimak Pass (Schumacher et al., 1982) and freshwater addition along the coast of Alaska. Apparently, the remaining volume comes across the shelf somewhere south of Cape Navarin.

Rogers (1981) examined relations between the North Pacific Oscillation (the wave-like nature of 700 mbar isobars which can be indexed in terms of opposition of sign of mean temperature anomaly between Alaska and western Canada) and Bering Sea ice. Niebauer (1983) also used 700-mbar charts to examine anomalies in percent ice cover. Although relationships exist between 700-mbar fluctuation and those in ice extent, it is more straightforward to consider ice-atmospheric interaction from the perspective of surface cyclone trajectories. During winter there is a tendency for two cyclone tracks, one parallel to the Aleutian Islands and one curving northward toward Siberia (e.g., Figure 7.2); but there always is a decrease in the number of cyclones with increasing latitude (Overland and Pease, 1982). The former track results in the time-averaged feature known as the Aleutian Low, while the latter fact results in the statistical feature called the Siberian High. The juxtaposition of these manifestations of fluctuations in 700-mbar isobar steerage of cyclones results in northeasterly mean winter winds over the eastern Bering Sea. These cold dry polar winds have an interannual signal which conditions water temperature and generates a strongly varying extent of ice cover.

The extent and magnitude of a given year's ice production impacts the abiotic environment in several ways: ice-melt generates local water column stability, horizontal gradients of density are formed, temperature of shelf waters (especially middle shelf bottom water) is established for the following summer, and local changes in flow and water properties occur where ice is produced. The first two phenomena are important to biota because water column stability is necessary for phytoplankton blooms associated with the marginal ice zone (e.g., Alexander and Niebauer, 1981), and plank-tonic life is at the whims of baroclinic transport. This latter feature was not clearly documented until winter 1983, but was found to result in mean along-marginal-ice-zone speeds of 5 to 8 cm/s during February through April over the shelf southwest of St. Matthew Island (Muench and Schumacher, 1984). Cold bottom water formed during winter persists through summer as a lower layer pool between the 50- and 100-m isobaths. The temperature of this layer in June has been corre-lated with degree days of frost for the preceding winter (Coachman and Charnell, 1979). In regions where ice is preferentially formed (e.g., the polynya south of St. Lawrence Island and Nunivak Island), brine is added which can be important to the regional salt budget and velocity field (Schumacher et al., 1983a).

Because meteorology in the eastern Bering Sea is storm dominated, vector mean winds are generally weak. Further, most of this shelf is outside the Rossby radius of deformation (a measure of the length scale over which coastal divergence significantly perturbs sea level slope). Away from the coast, currents respond as rotating vectors and little if any net

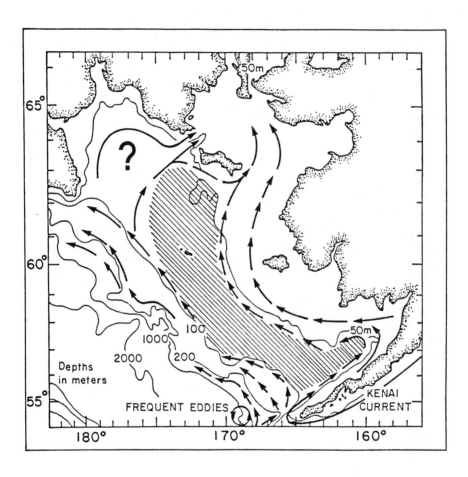

Figure 7.2. Schematic of long-term mean circulation based on direct observations, inferred baroclinic flow, water mass analyses, and model results (after Schumacher and Reed, 1983).

motion ensues (Schumacher et al., 1983b); however, turbulence due to wind waves and vertical shear play a key role in mixed layer depth. Sambrotto and Goering (1983) noted that the timing of storms is a critical factor which can either aid or detract from phytoplankton production as measured by nitrate uptake, and Incze (1983) showed considerable interannual variation in depth of the upper mixed layer in April, particularly over the middle shelf.

General circulation over the southeastern Bering Sea Shelf can be differentiated by regions related to water depth and forcing mechanisms (Schumacher and Kinder, 1983). Three low-frequency current regimes are present: the coastal (<50 m), the middle shelf (50 m – <100 m), and the outer shelf (>100 m). Coastal water from the Gulf of Alaska flow into the Bering Sea through Unimak Pass and continue northeastward along the Alaska Peninsula (Figure 7.2). This transport affects water properties, mass balance, and nutrients (Schumacher et al., 1982). The strongest baroclinic gradient is in the vicinity of the 50-m isobath (or associated with the inner front). The current flows counterclockwise around Bristol Bay and then northwest past Nunivak Island. As the magnitude of tidal currents decreases to the north (Pearson et al., 1981), both the inner front and coastal current appear to be focused near the 30-m isobath. Flow exits the system through Bering Strait. Current speeds averaged over periods greater than one month are statistically significant and vary from ~5 to 25 cm/s in Unimak Pass, and 1 to 6 cm/s along most of the path; the higher values are typically present in winter. Although interannual variation has not been documented in the mean along-isobath current, pulses of offshore transport occur along the Alaska Peninsula (Schumacher and Moen, 1983), and their magnitude likely varies from year-to-year.

Between the 50- to 100-m isobaths all data from the southeastern sub-region show statistically insignificant (<1 cm/sec) mean flow. Sluggish transport here is a year-to-year feature. When ice is not present, similar conditions appear to obtain for the more sparsely studied region between St. Lawrence Island and St. Matthew Island. In the vicinity of the middle front (or 100-m isobath) and over the outer shelf, current speeds are statistically significant with along-shelf means between 1 and 10 cm/s and cross-shelf means of 1 to 5 cm/s. The former current appears to be continuous over the outer shelf, flowing northwestward to the vicinity of Cape Navarin. Here, some portion of the flow follows the 80-m isobath toward the northeast, exiting the region through Anadyr Strait. Since the baroclinic component of the velocity field is important for along-isobath flow (Schumacher and Kinder, 1983), variations in quantity and extent of ice melt can result in inter-annual variability. For example, during winter 1983, currents in the vicinity of the 80- to 110-m isobaths southwest of St. Matthew Island were observed to be 5 to 10 cm/s, even over the "non-advective regime" of the middle shelf (Muench and Schumacher, 1984).

Currents over the slope, while poorly defined, have been suggested to impact outer shelf character-istics, particularly at long (greater than 10 day) periods. Kihara (1982) inferred changes in currents flowing northeastward along the Aleutian Islands and these correlate with changes in the mean bottom water temperature over the outer portion (>100 m) of the shelf. This suggests an interannual signal in trans-port of the Bering Slope Current. Off the southeastern shelf, the Bering Slope Current is rich in baroclinic structure and eddies but generally flows northwestward along the continental slope at 5 to 15 cm/s (Kinder et al., 1980; Paluszkiewicz and Niebauer, 1984). It is not clear where the Bering Slope Current originates, although water mass analysis suggests the presence of Alaskan Stream water. These waters likely entered the Bering Sea through the deeper passes in the western Aleutian Islands. All evidence suggests that slope transport is highly variable over time scales of months to years.

Hydrographic domains, which are nearly coincident with current regimes, can be delineated by depth and vertical structure (Figure 7.3). The shelf and oceanic domains are separated by a shelf-break front (Kinder and Coachman, 1978), which is ~50-km wide. Farther inshore is the middle front which lies approximately over the 80-100 m isobaths. This feature separates the three-layered stratification of the outer-shelf domain from the two-layered stratification of the middle-shelf domain and is structurally most pronounced in the lower water column (Coachman and Charnell, 1977, 1979). Inshore of this front, a third front parallels the 50 m isobath and separates the middle shelf from the unstratified waters of the coastal domain (Schumacher et al., 1979). Kinder and Schumacher (1981) summarized the hydrographic structure across the shelf and Coach-man et al., (1980) discussed the system of fronts. The frontal system is most clearly defined during summer, but it can be distinguished throughout the year using appropriate parameters.

DYNAMICS OF SELECTED SPECIES

We briefly examine the dynamics of four species which are of trophic as well as economic importance:

118

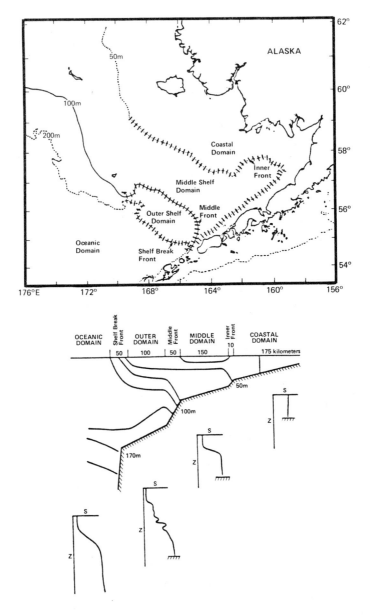

Figure 7.3. Approximate locations of hydrographic
domains and fronts (upper), and a schematic depiction
of a cross-shelf section (lower) with typical vertical
structure during periods of positive buoyancy input
(i.e., solar insolation and/or ice melt). The middle
domain is generally well mixed by late fall.

walleye (Alaska) pollock, Theragra chalcogramma; red
king crab, Paralithodes camtschatica; and two species
of Tanner crab, Chionoecetes bairdi and C. opilio. We
focus on prominent changes which reflect interannual
(year class) variability which may be related to interactions with major components of the biotic and abiotic
environment. The biology and fisheries of these
species are not reviewed here, but are provided in
other papers (Adams, 1979; Otto, 1981; Smith, 1981;
Francis and Bailey, 1983; Hayes, 1983; Incze et al.,
1984, in press; Lynde, 1984).

Pollock

Commercial quantities of pollock have been harvested from the eastern Bering Sea since 1964. Since
1970, this species has contributed from 73% to 85% of
the annual commercial harvest of all major fisheries of
this region other than salmon. From 1970 to 1976,
annual harvests of pollock ranged from 1.25 to 1.87
million metric tons. Restrictions (catch quotas) were
placed on this harvest after 1976 because of a significant decrease in catch per unit effort over the previous 4 yr. Since 1977, the harvest of pollock has
remained quite stable at approximately 960,000 metric
tons (10^3 kg) yr^{-1}, with a coefficient of variation
(CV) of only 2.6%. However, this recent stability in
harvests belies considerable variability in year-class
abundance. Considering that pollock are fully recruited to the fishery at a young age (mostly by 3 yr), the
apparent stability in harvests is probably due in part
to the vast size of the eastern Bering Sea ecosystem.
Francis and Bailey (1983) estimated by cohort
analysis that recruitment at age 1 between 1974 and
1980 varied by a factor of 7, from approximately 5 to
35 billion individuals, with a CV of 44% for the
population of the eastern Bering Sea shelf. When
separate analyses were conducted on data from catches
northwest and southeast of the Pribilof Islands, the CV
of recruitment in the southeast was from 3 to 6 times
as large as the northwest, indicating greater variability in year-class abundance over the southeastern
shelf. The growth (size at age) characteristics of
fish from northwest and southeast of the Pribilofs are
distinct (Lynde, 1984), suggesting that at least two
"production units" of the stock exist in the eastern
Bering Sea. From 1973 to 1977, 62% of the pollock
harvested from shelf came from northwest of the
Pribilofs, while 75% came from southeast during the
period 1981-1982. Indeed, cohort analyses (Francis and
Bailey, 1983) indicate that these regional patterns in
harvest resulted from particularly abundant year
classes originating in each region in 1973 and 1978,

respectively. The 1972 year class appears to have been quite successful in both regions though data from this year are somewhat equivocal (R. Francis, personal communication). Since each year class is in the fishery for at least 2 yr, and strong year classes may last longer (Bakkala, 1983), it has been possible for several years to harvest fairly stable amounts of pollock from the eastern Bering Sea as a whole, even though the area of production has varied considerably. Since pollock constitute such a substantial portion of total fisheries yield from this region, this variability may assume considerable importance. We discuss mechanisms for geographical and interannual differences in year-class abundance in a later section.

Red King Crab

Red king crab in the eastern Bering Sea is among the best sampled fisheries populations anywhere. Bottom trawls have been used to survey the population of pre-recruits (at least one year before entering the commercial fishery) each year since 1967 (Hayes, 1983). Crabs are fully recruited to the sampling gear at about age 6, but reliable indices of abundance (as opposed to estimate of abundance) are provided for 4-yr-old crab as well (Incze et al., in press). The harvested biomass of king crab listed in Table 7.1 represents catch at a very low level of the stock (see Figure 7.4). In 1983 the fishery was entirely closed as a precaution against reducing the reproducing stock to levels from which recovery might be very slow.

Hayes (1983) used Japanese and Russian data on catch per unit effort (CPUE) and scaled it to estimates of abundance made from U.S. trawl surveys. The indexed CPUE data provide estimates of changes in stock abundance prior to 1967 and indicate a high level during the late 1950s and early 1960s. However, these data probably are affected by low levels of effort prior to 1960 and by rapidly increasing effort thereafter (Otto, 1981:1044), making it difficult to accurately describe the previous "cycle" in king crab abundance. CPUE decreased dramatically during the middle and late 1960s, similar to the landings depicted in Figure 7.4.

In an analysis of year class abundance of 4 yr olds from 1968 to 1983, Incze et al., (in press) found that the largest index of year-class size was 11.5 times the smallest. They identified three strong year classes: 1971, 1972, and 1978 (Figure 7.4). The first two arose from comparatively small stocks of reproducing female crabs indicating an extremely high survival rate for the progeny compared to other years. These strong year classes dominated the fishery in the late

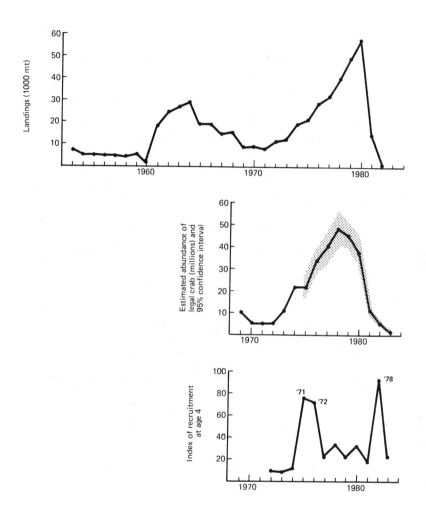

Figure 7.4. Historical catch of red king crab from the
eastern Bering Sea (top), changes in estimated
abundance of legal-sized crab from surveys with 95%
confidence intervals beginning with 1975 (middle), and
changes in year-class strength assessed at age 4
(bottom). Historical catch data are from Otto (1981)
and National Marine Fisheries Service, Seattle,
(unpubl.); survey estimates and 4-yr-old indices are
from Incze et al. (in press). Year classes are
identified as 1971, 1972, 1978 in the bottom figure.

1970s (Figure 7.4). The 1978 year class originated from a very large reproducing stock, but the actual success ratio appeared comparatively unremarkable (index of recruitment success = index of 4-yr-old abundance divided by estimated abundance of adult female crabs producing the year class). These results show dramatic differences in year-class abundance produced by age 4, though they do not identify more precisely when in the life history of these crabs these differences were established. The findings further indicate both positive and negative relationships with stock size. While differences in abundance were clearly evident by age 4, and therefore important in understanding variability in production at the level of the resource, success of the early life history stages did not always establish the level of later harvests. The 1971 and 1972 year classes went on to support large, prosperous harvests, while the 1978 year class failed to show up in significant numbers as 5 yr olds. Major influences on fisheries recruitment therefore seem to be possible over a long time period in the growth of this species.

Tanner Crabs

Tanner crabs, which inhabit broad regions of the shelf (Figure 7.5), are ecologically significant because of their great numbers and their position high in the benthic food web (Feder and Jewett, 1981). They have, in addition, proven to be a valuable resource, particularly in recent years with the decline of king crab stocks. Tanner crabs are considerably smaller than king crabs and are not sampled as well by the trawl gear used in surveys. Variability in their growth rates and the older age at recruitment to the survey gear make it difficult to distinguish size modes from the size (carapace width) frequency data which are available from such surveys. Year classes, therefore, are not regularly identifiable, and general trends in abundance must be used to infer variability in recruitment. Somerton (1981) used size-frequency data to show that strong year classes appeared to occur often in C. bairdi than C. opilio, and that the latter species recruited more regularly to the north than to the south (see species ranges in Figure 7.5). Subsequent surveys also have shown much greater decline in C. opilio stocks than C. bairdi, which may have been due to the prominence of a few strong year classes of C. opilio, now leaving the population through mortality. A 5-yr study of larval abundance patterns (Incze, 1983) indicated consistently strong larval year classes for C. bairdi over the most of the area of study (southeastern shelf), but only occasionally strong ones for

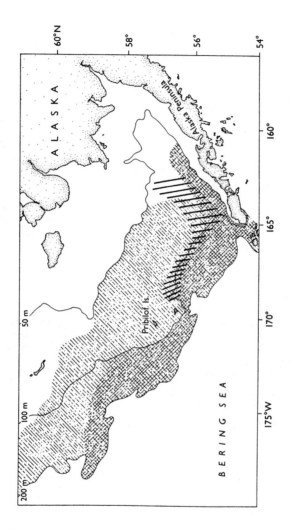

Figure 7.5. Average (1975 to present) distribution of major stocks of Chionoecetes opilio (light shading) and C. bairdi (dark shading) in the eastern Bering Sea. Considerable overlapping of the two species occurs during some years in the middle shelf domain of the southeastern shelf in the area marked by cross-hatchings. C. bairdi extends south and east into the Gulf of Alaska and west along the Aleutian Island chain; C. opilio is distributed further north and along the western Pacific margin south to Japan.

C. opilio. In particular, comparison of data for 1978, 1980, and 1981 (the most thoroughly studied years) showed three orders of magnitude change in abundance of larval C. opilio. Over part of the shelf, differences in larval abundance could be accounted for by rapid declines in reproducing stock, probably due to aging and mortality of earlier successful year classes.

However, this did not explain changes observed over about half of the study area where larval survival appeared to be the most important factor (Incze et al., in prep.). Larval survival was indexed as abundance compared to size of the spawning stock. Possible explanations for these changes are briefly discussed in the following section. As with king crab, what appeared to be a successful start to a year class (based on larvae in 1978 in this case) has not led to a large year class entering the fishery. This finding is also addressed below.

DISCUSSION

Numerous factors may act independently or interactively to affect the abundance of a year class of organisms, and various factors may assume different significances from year-to-year. It has been suggested that this complexity will ultimately prevent complete understanding or predictability of year-class size. Whether or not this is so, considerable gain is possible if certain phenomena can be identified as being particularly influential for a given species and environment. For each of the stocks mentioned above there are some relationships with the biotic and abiotic environment which, from our present state of knowledge, appear likely to be important. It would be premature to offer any hypotheses based on these relationships, yet they provide insights into processes which may be affecting the abundance of important living resources. Since these resources probably will be harvested and studied for many years into the future, we anticipate progress in the formation and testing of ideas related to recruitment of these species.

Pollock spawn during an on-shelf migration which begins during early March over the slope and continues through June into shallower shelf waters. Pollock migrate into the middle shelf domain (Figures 7.3, and 7.6) when bottom water temperature is suitable, but appear to avoid this region when temperatures are below 2.5°C (Nishiyama and Haryu, 1981; Francis and Bailey, 1983). Temperature therefore appears to be a major determinant of spawning distribution; strong interannual differences in bottom water temperature over the middle shelf (see earlier review of the physical

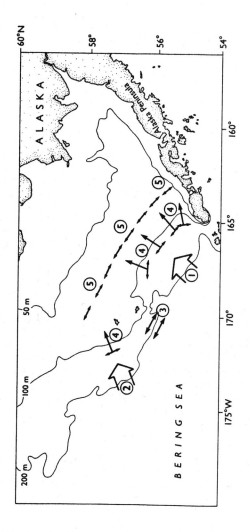

Figure 7.6. Possible migratory paths of pollock during
the spawning period. (1) and (2): Shelfward migration
from slope to outer shelf during March. (3) Possible
along-shelf migration during the early phase of on-
shelf movement. (4) Migration into the middle shelf
domain if conditions are suitable; extent and
directional preference of movement over outer shelf if
there is no significant migration to the middle shelf
is unknown. (5) Possible along-shelf movement of
pollock after entering the middle shelf domain.

environment) should cause greatly varying spatial patterns of spawning effort (Figure 7.6) (cf. Francis and Bailey, 1983).

Differences in transport over the outer and middle shelf domains of the southeastern Bering Sea should systematically affect the distribution of eggs and larvae after spawning. Upper layer mean flow of the outer shelf carries eggs and larvae to the northwest, roughly parallel to the isobaths, at speeds which range from approximately 1 to 10 cm/s, or roughly 60-600 km during the period of egg and larval development. The underlying causes of variability in this flow rate are not well known, but probably involve oceanic forcing and baroclinic flow induced by melting sea ice. The upper range of flow (anywhere from 6-10 cm/s) sustained over the period of planktonic egg and larval develop-ment is sufficient to remove these stages to the north-west, while periods of low flow should enable eggs and larvae to remain over the southeastern outer shelf. Horizontal mixing of water between the outer shelf, slope and oceanic domains may be significant early in the spawning period, such as during March (Vidal and Smith, in press). Eggs spawned over the middle shelf are expected to remain there due to weak circulation. If the geographic location at the time of transition to juvenile stage determines geographic association of shelf and slope populations in future years, as the studies of Lynde (1984) suggest, the variability of the middle shelf bottom water temperature and outer shelf transport may be involved in the dynamics of two separate production units suggested by the analyses of Francis and Bailey (1983). Muench and Schumacher (1984) demonstrated that low temperature and increased baroclinic transport (both related to the melting sea ice) occur together. Transport due to other factors, however, may be unrelated to temperature.

In any of the above cases, the actual abundance of fish in any year class and location is determined by factors such as larval feeding success and predation on larval and juvenile fish (Paul, 1983; Incze et al., 1984; Lynde, 1984; Smith et al., 1984; Dwyer et al., in press). Variability of larval feeding success has not been documented for pollock in the eastern Bering Sea, but variability of the potential prey field has. Walsh (1983) reported an order of magnitude difference in biomass of small net zooplankton (mostly Pseudocalanus spp.) over the middle shelf during May of 1979 and 1980, and Incze et al. (in prep.) found similar differ-ences in abundance number m^{-2} of Pseudocalanus spp. between April 1978 and April-early May of 1980 and 1981. Smith and Vidal (in press) showed significant differences in production of Pseudocalanus spp. (changes in abundance of stage 3 and 4 copepodites) between May of 1980 and 1981. Whether these differences in abundance of a primary prey (see Incze et al., 1984) were sufficient to result in significant interannual differences in growth and survival of larval pollock is unknown. The 1978 year class of pollock from the southeastern shelf was large (Francis and Bailey, 1983), but year classes originating in 1979, 1980, and 1981 appear to be weak (National Marine Fisheries Service, Seattle, Washington, unpubl. data). Comparison of other conditions in 1978 and 1979 would be particularly instructive because of the apparent similarities in abundance of a primary prey. However, the distribution of spawning effort, reasonably well known for 1978, is poorly documented for 1979 (Figures 1 and 2 of Incze et al., 1984). Consequently, we have little information with which to speculate on the cause(s) of differences in strength of the two year classes. Similar problems also pertain to any consideration of the 1980 and 1981 year classes. Pseudocalanus spp. were significantly less abundant during springs of 1980 and 1981 than in the previous two years over the middle shelf, but spawning distribution was not well recorded. As explained below, other potentially important variables also were not sampled.

Little field information is available regarding predation on larval pollock. Clupeid, osmerid and ammodytid fish are likely predators based on their size, feeding ecology and general distributions (Smith et al., 1984), but predation rates have not been examined. Gelatinous zooplankton could be important and might show considerable interannual variations in abundance, timing or distribution, but these organisms are not well studied in this region. Avian predators may be important for juvenile pollock (Hunt et al., 1981) but probably not for larvae (Hunt, et al., unpub. data).

Predation on juvenile pollock by other fish is known to be intense (Smith et al., 1984). Cannibalism by older pollock (Takahashi and Yamaguchi, 1972; Bailey and Dunn, 1979; Dwyer et al., in press) and predation by cod (Gadus macrocephalus: Dunn, 1979) may be responsible for the most significant predatory losses of young fish. While predation on juveniles may be a significant source of mortality, it must vary between years to be a major cause of variations in recruitment to adult populations or to the fishery. Francis and Bailey (1983) showed that vertical separation of juveniles (ages 0 and 1) from older pollock can exist while the water column is stratified (older fish are deeper), but might not occur in a mixed (isopycnal) water column. These authors suggest that interannual differences in the time of autumnal mixing (due to surface cooling and wind stress) may significantly affect losses due to cannibalism. Stratification which persists later into the autumn presumably promotes additional growth of juvenile fish due to warmer temperatures and possible better feeding conditions, so that when mixing does occur, juveniles are larger and may be less susceptible to predation by older fish. Adult pollock appear to leave the middle shelf during winter (Serobaba, 1968; A. M. Shimada, National Marine Fisheries Service, Seattle, Washington, unpubl. data) so an extension of stratified conditions later into the fall, even by two to three weeks, may result in a considerable proportional decrease in mortality. Variations in the extent of onshelf migration of pollock during spawning, and therefore the distribution of O-age fish, also could be influential because shallower water mixes earlier in the autumn than deeper water. Other variables include the distribution and abundance of cod and possible variations in the time at which older pollock leave the middle shelf.

Red king crab are broadly distributed during winter, primarily over the middle shelf domain, and undergo a southeastward migration toward the Alaska Peninsula late in winter for mating (Hayes, 1983; Weber, 1974 and references cited therein). Egg hatching occurs prior to mating, mostly during April and May (Takeuchi, 1962; INPFC, 1963; Weber, 1967; Haynes, 1974; D. Armstrong et al., University of Washington, Seattle, unpubl. data). The shelf depth at which the majority of female crabs hatch their eggs is not known, but larval data from 1982 indicate that the majority of hatching occurred deeper than 40 m (D. Armstrong, unpubl. data). This distribution would place many larvae within the influence of the along-shelf flow associated with the 50-m isobath (Figure 7.7). The distribution of larvae prior to benthic settlement (e.g., Hebard, 1961; Haynes, 1974) would be influenced by the extent of association with the 50-m current, variability in the velocity of this current, and cross-shelf (across-isobath) perturbations in flow. The ultimate spatial distribution of newly-settled crabs of a year class also would be a function of the east-west distribution of the spawning population, something which has undergone significant changes (in an eastward direction) during the past few years (Incze, in prep.). The area of settlement may have an influence on post-larval survival due to differences in substrate type (refuge and food) and abundance of predators. Yellowfin sole (Limanda aspera) can be extremely abundant over the middle shelf (Bakkala, 1981) and is a known predator of megalops stage king crab (K. Haflinger, University of Alaska, Fairbanks, unpubl. data) and possible early benthic instars as well. Spatial overlap of this predator with settling king crab larvae,

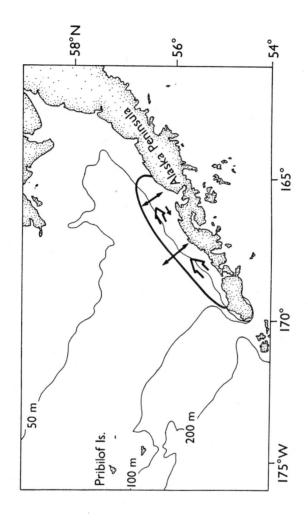

Figure 7.7. General distribution of major stocks of female red king crab during the larval hatching period. Large, open arrows show the residual circulation associated with the 50-m isobath; small arrows show possible perturbations to this flow which should alter the depth of settlement of the first benthic stage as well as the northeasterly distance that larvae are transported from the hatching location. Reversals of mean flow have been observed landward of 50 m.

particularly in areas affording little spatial refuge for crabs, may be a significant problem for year-class survival. Yellowfin sole were in great abundance during the late 1970s, and considerable variations in catch per unit effort from survey vessels has been reported for the red king crab grounds (Bakkala and Wespestad, 1983). Temperature seems to affect the distribution of this species, as it appears to migrate on- and off-shelf during summer and winter, respectively, and may avoid bottom water of less than 0°C (Wakabayashi, 1974; but also see Bakkala, 1981). It is conceivable that cold bottom water temperatures over the middle shelf during some years (Figure 7.8) deters migration of yellowfin sole into this region and keeps them more concentrated elsewhere. Such a response would reduce spatial overlap of yellowfin sole with C. opilio, but possibly increase interactions with C. bairdi and red king crab north of the Alaska Peninsula. Other flatfishes also may be important depending on composition of their diets and changes in abundance.

Year-class abundance of red king crab has shown both positive and independent responses to spawning stock size (Incze et al., in press). The comparatively spectacular success of the 1971 and 1972 year classes, at a time when adult stocks were very low, provide an opportunity to investigate major climatic conditions which may have been different (for instance, alongshore and offshore components of wind-generated transport during the period of larval development [Schumacher and Incze, unpubl. data]). Biotic changes, such as the distribution and abundance of major predators also must be considered. The existing data for this period do not permit a detailed examination of the possible mechanisms for the high success rate, but the environmental relationships may lead to hypotheses which can be tested and reformulated as new data are acquired.

With the 1978 year class there was a large abundance of 4 yr olds that resulted from a very large population of adult crab (therefore no over-riding density dependent negative feedback up to this age), but these young crabs did not survive in high abundance to the following year. In fact, larger crabs (older year classes) also experienced a significant decrease in abundance from 1982 to 1983. One possibility is that fish from the record 1977 year class of cod (Gadus macrocephalus) became large enough to consume king crab of a wide range of sizes (approximately 60-cm fork length) (J. June, National Marine Fisheries Service, Seattle, Washington, personal communication; see Figure 7.9). The 1978 year class of crab in particular may have been subjected to significant predation by cod for the first time during 1982-1983 because king crab disperse from juvenile aggregations after their third

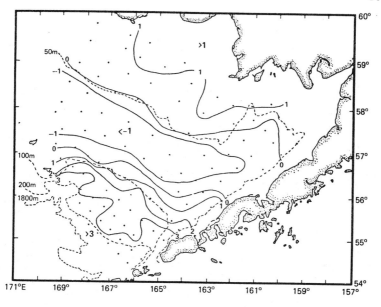

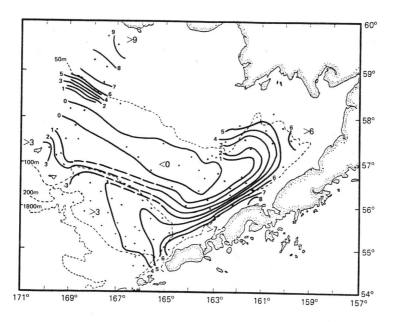

Figure 7.8.   Bottom water temperature (°C) during a
cold year (1976):   June (upper, from Kinder, 1977) and
late September-early October (lower: after Kinder and
Schumacher, 1981).   Note persistence of cold temperatures over the middle shelf throughout summer.

132

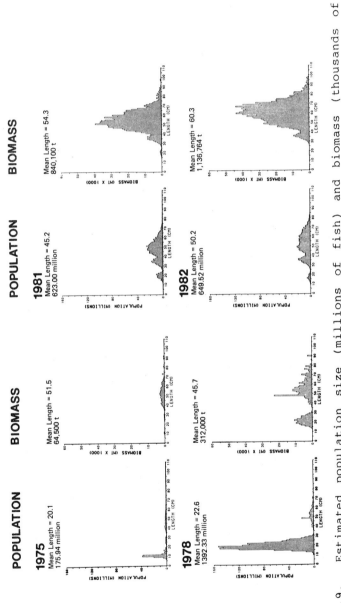

Figure 7.9. Estimated population size (millions of fish) and biomass (thousands of metric tons) of cod (Gadus macrocephalus) related to size (fork length) of the fish. Mean length and total numbers and biomass are listed at top of each figure. Note size and longevity of the 1977 year class evident in the prominent mode in data from 1978 onward (data from National Marine Fisheries Service, Seattle, Washington).

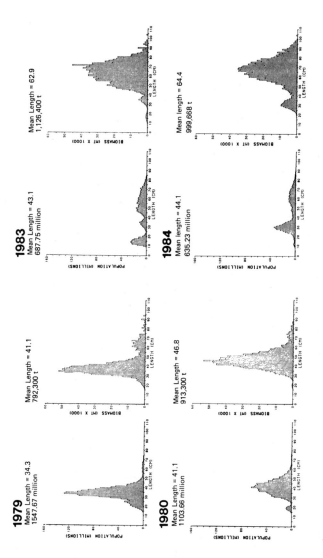

Figure 7.9.   Continued.

year (Incze et al., in press) and therefore may be more
subject to predation by widely dispersed predators.
However, the apparent timing of the great losses,
during winter (R. S. Otto, National Marine Fisheries
Service, Kodiak, Alaska, unpubl. data), suggests that
other causes, possibly disease also were important
(R. S. Otto, personal communication).

The signiticant finding thus far is that recruitment of red king crab to the fishery appears to be a
function of events during early life (as evidenced by
three year classes) and also fairly late in juvenile
life (as evidenced for 4- to 5-yr-old crabs of the 1978
year class). It is not known how frequently the latter
case is important.

Tanner crabs do not appear to undertake significant regular migrations, and egg hatching occurs over
broad regions where adult populations are distributed
(Incze, 1983; also see Figure 7.5). Both species have
highly synchronized and pulsed periods of hatching:
early to middle April for C. opilio, and early to
middle May for C. bairdi (Incze et al., 1982). Though
some interences can be made about recruitment variability based on spatial and temporal patterns of adult
distribution and abundance, some of the most convincing
data demonstrating interannual differences in potential
year-class strength come from studies of the larval
stages. Larval C. opilio have shown temporal changes
in abundance which, when related to the abundance of
adult female crabs, indicated significant interannual
differences in survival (Incze et al., in prep.). The
greater variability of larval survival of C. opilio
compared to its congener may be related to its earlier
hatching and its habitation of a more variable region
of the shelf, though the linkages remain unknown. Poor
larval survival (indexed to the size of the spawning
stock) coincided with years of (1) low abundance of
Pseudocalanus sp. copepods and (2) deeper upper mixed
layers during April, when C. opilio larvae hatch.
Pseudocalanus sp. are a numerically dominant copepod of
the middle shelf (Vidal and Smith, in press) and probably are a major prey of Tanner crab zoeae (Paul et
al., 1979; Incze, 1983; Incze and Paul, 1983). These
copepods were an order of magnitude more abundant in
the one year of study in which larval survival was very
high, suggesting that feeding conditions influenced the
degree of larval success (Incze et al., in prep.).
Interannual differences in abundance of Pseudocalanus
spp. remain unexplained because of the small number of
years of data, but may involve temperature (Vidal,
1980; Smith and Vidal, in press). Predation does not
appear to have been an important factor affecting
abundance of Pseudocalanus spp. during springs of 1980
and 1981 (Smith and Vidal, in press).

Although early larval stages of C. opilio showed
very high survival rates through their first two months
during 1978, these crabs have not yet appeared in the
surveys, as they should have based on rates of growth
and size of recruitment to the survey gear. There is
evidence that predation on the megalops and early
instar benthic crabs by yellowfin sole is a significant
problem over the middle shelf (K. Haflinger, unpubl.
data) and may have been responsible for cropping an
otherwise successful year class. The direct effects of
environmental conditions on migrations, as well as
changes in year class size, may vary the extent of predation on C. opilio by yellowfin sole. There also is
evidence that Tanner crabs suffer high predation from
cod (J. June, unpubl. data), which have dramatically
increased in numbers and biomass as a result of the
very successful 1977 year class (Figure 7.9). Cod
begin taking significant numbers of Tanner crab when
the fish reach 30-40 cm fork length (J. June, unpubl.
data) and could have begun affecting the Tanner crab
population as early as 1980. Possible interannual
variations in the migrations of cod also should be considered. As with red king crab, factors in early life
history and later life both appear to be important in
accounting for strength of recruitment to the fishery.
    Studies indicate that early larval survival of C.
bairdi may not be as variable as C. opilio, so this
potential source of recruitment variability may be
lower for this species. However, the time period over
which observations have been made is still very brief.
The position of the two stocks with respect to water
masses and current regimes suggests that stocks of C.
opilio in the southeastern Bering Sea (where they are
at the periphery of their range) may require more time
to recover from depressed levels because currents do
not favor dispersal of larvae into this area from
regions of sustained high abundance farther north. The
population therefore may be marked by the presence of
occasionally strong year classes. One result, in addition to highly variable harvestable stocks, is extreme
variations in size of the spawning stock. Currents
favor dispersal of larval C. bairdi to the periphery of
this species' range (cf. Figures 7.2 and 7.5), so that
recovery of the stock at its margins may take place
more rapidly. Larvae of C. bairdi from the outer shelf
may experience some of the variability in northwestward
flow experienced by larval pollock, but will do so beginning later in the season. Predation on larval (K.
Haflinger, unpubl. data) and benthic stages (Jewett,
1978; J. June, unpubl. data) of C. bairdi may be
considerable, but thus far in the record there have
not appeared to be very strong interannual fluctuations
in recruitment.

It is interesting to note that 1978 produced extremely abundant year classes of at least three species in the southeastern Bering Sea: (1) C. opilio, as determined at the larval stage, (2) red king crab, as determined at age 4; and (3) pollock, as determined at ages 3 and 4. Year classes of all three species were much more successful in 1978 than in neighboring years. Even though the 1978 year class of king crab was, in fact, produced by a large spawning stock, this same stock had failed to produce a large year class in 1977, when it might first have done so. Whether it spawned a large 1979 year class of larvae or other early life stages is not known because the survey would not have sampled these crabs reliably until their fourth year, 1983. There certainly was no large 1979 year class sample at this time, but whether this was due to the same factor(s) that affected the rest of the population during 1982-1983 or to earlier events cannot be ascertained.

In the above discussion, we have commented briefly on what probably are complex processes governing recruitment. For a small number of species, we have stressed interactions with good potential for inter-annual changes and for which we have some data. There are many interactions we cannot presently quantify, and the mechanisms we have suggested clearly address a limited set of possibilities. For instance, our discussion certainly prompts one to ask how the 1977 year class of cod arose. Yet we have little information on the winter period when this species spawns its demersal eggs. Nor do we know much about the spawning grounds or the requirements of its pelagic larvae (Gunderson, 1983). Also, we know little about the distribution of juvenile cod and pollock during winter, or the timing of cross shelf movements of adult fish, particularly during autumn. Likewise, the causes of good and poor larval year classes of pollock and Tanner crabs, which are only irregularly sampled, can only be guessed. Despite many unknowns, the strong biotic and abiotic fluctuations that occur in the eastern Bering Sea indicate that much can be learned, and these strong signals suggest relationships worthy of further investigation.

ACKNOWLEDGMENTS

We thank many colleagues for discussions instrumental to this work, especially Drs. Kevin Bailey, Lawrence Coachman, Robert Francis, Murray Hayes, Robert Otto and Joe Niebauer. We also are grateful to Richard Bakkala and Jeff June for their help with fisheries

data from the National Marine Fisheries Service. Preparation of this manuscript was partially supported by
the NOAA Administrator's Discretionary Funds.

NOTES

1. Contribution number 689 from the School of
Fisheries, University of Washington and number 709 from
the Pacific Environmental Laboratory, Seattle.

REFERENCES

Adams, A. E. 1979. The life history of the snow crab,
Chionoecetes opilio: A literature review. Alaska
Sea Grant Rep. 78-13, Univ. Alaska, Fairbanks, 141
pp.
Alexander, V., and Niebauer, H. J. 1981. Oceanography
of the eastern Bering Sea ice edge zone in spring.
Limnol. Oceanogr., 16:1111-1125.
Armstrong, D. [1984]. Univ. Washington, Seattle,
Unpub. data.
Bailey, K. M., and Dunn, J. 1979. Spring and summer
foods of walleye pollock, Theragra chalcogramma,
in the eastern Bering Sea. Fish. Bull., U.S.
77:304-308.
Bakkala, R. G. 1981. Population characteristics and
ecology of yellowfin sole. In The eastern Bering
shelf: Oceanography and resources, pp. 553-574.
Ed. by D. W. Hood and J. A. Calder. Univ. Wash-
ington Press, Seattle, 1339 pp.
Bakkala, R. G. 1983. Other species. In Condition of
groundfish resources of the eastern Bering Sea and
Aleutian Islands Region in 1982, pp. 181-187.
NOAA Tech. Memo. NMFS F/NWC-4, 187 pp.
Bakkala, R. G., and Wespestad, V. 1983. Yellowfin
sole. In Condition of groundfish resources of
the eastern Bering Sea and Aleutian Islands Region
in 1983, pp. 42-68. Ed. by R. G. Bakkala and
L. L. Low. Unpubl. Rep., Northwest and Alaska
Fisheries Center, Nat. Mar. Fish. Serv., Seattle,
Washington, 154 pp.
Coachman, L. K., and Aagaard, K. 1981. Re-evaluation
of water transports in the vicinity of Bering
Strait. In The eastern Bering Sea shelf:
Oceanography and resources, Volume 1, pp. 95-
110. Ed. By D. W. Hood and J. A. Calder. Univ.
Washington Press, Seattle, 1339 pp.
Coachman, L. K., and Charnell, R. L. 1977. Finestruc-
ture in outer Bristol Bay, Alaska. Deep-Sea Res.
24:869-889.
Coachman, L. K., and Charnell, R. L. 1979. On lateral
water mass interaction--A case study, Bristol Bay,
Alaska. J. Phys. Oceanogr. 9:278-297.

138

Coachman, L. K., and Kinder, T. H., Schumacher, J. D., and Tripp, R. B. 1980. Frontal systems of the southeastern Bering Sea Shelf. In Stratified flows, pp. 917-933. Ed. by T. Carstens and T. McClimans. Sec. Int. Assoc. Hydraulic Res. Symposium, Trondheim, June 1980, TAPIR Publishers, Norway.

Cooney, R. T., and Coyle, K. O. 1982. Trophic implications of cross-shelf copepod distributions in the southeastern Bering Sea. Mar. Biol. 70:187-196.

Dunn, J. 1979. Predator-prey systems in the eastern Bering Sea. In Predator-prey systems in fisheries management, pp. 81-91. Sport Fish. Inst., Washington, D.C.

Dwyer, D. W., Bailey, K., Livingston, P., and Yand, M. In press. Some preliminary observations on the feeding habits of walleye pollock (Theragra chalcogramma) in the eastern Bering Sea, based on field and laboratory studies. Bull. Int. North Pac. Fish. Comm.

Feder, H. M., and Jewett, S. C. 1981. Feeding interactions in the eastern Bering Sea with emphasis on the benthos. In The eastern Bering Sea shelf: Oceanography and resources, pp. 1229-1262. Ed. by D. W. Hood and J. A. Calder. Univ. Washington Press, Seattle, 1339 pp.

Francis, R. C. [1984]. Personal communication.

Francis, R. C., and Bailey, K. M. 1983. Factors affecting recruitment of selected gadoids in the northeast Pacific and eastern Bering Sea. In From year to year: Interannual variability of the environment and fisheries of the Gulf of Alaska and the eastern Bering Sea, pp. 35-60. Ed. By W. S. Wooster. Washington Sea Grant Publ. No. WSG-WO 83-3, Seattle, 208 pp.

Gunderson, D. R. 1983. Interannual variability of the environment and gadoid fisheries of the Gulf of Alaska and eastern Bering Sea. In From year to year: Interannual variability of the environment and fisheries of the Gulf of Alaska and eastern Bering Sea, pp. 61-69. Ed. by W. S. Wooster. Washington Sea Grant Publ., No. WSG-WO 83-3, Seattle, 208 pp.

Haflinger, K. [1984]. Univ. Alaska, Fairbanks, Unpubl. data.

Hayes, M. L. 1983. Variation in the abundance of crab and shrimp and some hypotheses on its relationship to environment causes. In From year to year: Interannual variability of the environment and fisheries of the Gulf of Alaska and the eastern Bering Sea, pp. 86-101. Ed. by W. S. Wooster.

Washington Sea Grant Publ. No. WSG-WO 83-3, Seattle, 208 pp.

Haynes, E. B. 1974. Distribution and relative abundance of larvae of king crab, Paralithodes camtschatica, in the southeastern Bering Sea. Fish. Bull., U.S. 72:804-812.

Hebard, J. F. 1961. Currents in southeastern Bering Sea. Int. North Pac. Fish. Comm. Bull. 5:9-16.

Hunt, G. L., Jr. [1984]. Unpubl. data.

Hunt, G. L., Jr., Burgeson, B., and Sanger, G. 1981. Feeding ecology of seabirds of the eastern Bering Sea. In The eastern Bering Sea shelf: Oceanography and resources. pp. 625-648. Ed. by D. W. Hood and J. A. Calder. Univ. Washington Press, Seattle, 1339 pp.

Incze, L. S. 1983. Larval life history of Tanner crabs, Chionoecetes bairdi and C. opilio, in the southeastern Bering Sea and relationships to regional oceanography. Dissertation, Univ. of Washington, Seattle, 191 pp.

Incze, L. S., and Paul, A. J. 1983. Grazing and predation as related to energy needs of stage I zoeae of the Tanner crab Chionoecetes bairdi Brachyura, Majidae). Biol. Bull. 165:197-208.

Incze, L. S., Armstrong, D. A., and Wencker, D. L. 1982. Rates of development and growth of larvae of Chionoecetes bairdi and C. opilio in the southeastern Bering Sea. In Proc. First International Symposium on the Genus Chionoecetes, pp. 191-218. Alaska Sea Grant Rep. No. 82-10, Univ. Alaska, Fairbanks, 792 pp.

Incze, L. S., Clarke, M. E., Goering, J. J., Nishiyama, T., and Paul, A. J. 1984. Eggs and larvae of walleye pollock and relationships to the planktonic environment. In Proc. Workshop on walleye pollock and its ecosystem in the eastern Bering Sea, pp. 109-159. Ed. By D. H. Ito. NOAA/NMFS Tech. Mem. NMFS F/NWC-62, 292 pp.

Incze, L. S., Otto, R. S., and McDowell, M. K. In press. Variability of year class strength of juvenile red king crab, Paralithodes camtschatica, in the southeastern Bering Sea. Can. J. Fish. Aquat. Sci., Spec. Publ.

INPFC, (International North Pacific Fishery Commission). 1963. Research by Japan. II. King crab investigations in the eastern Bering Sea. Int. North. Pac. Fish. Comm. Ann. Rep. 1963:88-101.

Iverson, R. L., Coachman, L. K., Cooney, R. T., English, T. S., Goering, J. J., Hunt, G. L., Macauley, M. C., McRoy, C. P., Reeburghand, W. S., Whitledge, T. E. 1979. Ecological significance of fronts in the southeastern Bering Sea. In

Coastal ecological processes, pp. 437-468. Ed. by R. J. Livingston, Plenum Press, New York, 614 pp.

Jewett, S. C. 1978. Summer food of the Pacific cod, Gadus macrocephalus, near Kodiak Island, Alaska. Fish Bull., U.S. 76:700-706.

Johnson, A. G. 1976. Electrophoretic evidence of hybrid snow crab, Chionoecetes bairde x opilio. Fish. Bull., U.S. 74:693-694.

June, J. [1984]. Northwest and Alaska Fisheries Center, Nat. Mar. Fish. Serv., Seattle, Washington, Personal communication.

Kihara, K. 1982. Fluctuations of the water temperature and the salinity in the eastern Bering Sea. Bull. Jpn. Soc. Sci. Fish. 48:1685-1688.

Kinder, T. H. 1977. The hydrographic structure over the continental shelf near Bristol Bay, Alaska, June 1976. Tech. Rep. m77-3, Dep. Oceanography, Univ. Washington, Seattle, 61 pp.

Kinder, T. H., and Coachman, L. K. 1978. The front overlaying the continental slope of the eastern Bering Sea. J. Geophys. Res. 83:4551-4559.

Kinder, T. H., and Schumacher, J. D. 1981. Hydrography over the continental shelf of the southeastern Bering Sea. In The eastern Bering Sea shelf: Oceanography and resources, pp. 31-52. Ed. by D. W. Hood and J. A. Calder. Univ. Washington Press, Seattle, 1339 pp.

Kinder, T. H., Schumacher, J. D., and Hansen, D. V. 1980. Observation of a baroclinic eddy: an example of mesoscale variability in the Bering Sea. J. Phys. Oceanogr. 10:1228-1245.

Lynde, C. M. 1984. Juvenile and adult walleye pollock of the eastern Bering Sea: Literature review and results of ecosystem workshop. In Proc. Workshop on walleye pollock and its ecosystem in the eastern Bering Sea, pp. 43-108. Ed. by D. H. Ito. NOAA Tech. Memo. NMFS F/NWC-62, 292 pp.

Muench, R. D., and Schumacher, J. D. 1984. On the Bering Sea ice edge front. J. Geophys. Res. (In press).

National Marine Fisheries Service. [1984]. Northwest and Alaska Fisheries Center, Seattle, WA, Unpubl. data

Niebauer, H. J. 1980. Sea ice and temperature fluctuations in the eastern Bering Sea and the relationship to meteorological fluctuations. J. Geophys. Res. 85:7507-7515.

Niebauer, H. J. 1983. Multiyear sea ice variability in the eastern Bering Sea: an update. J. Geophys. Res. 88:2733-2742.

Nishiyama, T., and Haryu, T. 1981. Distribution of walleye pollock eggs in the uppermost layer of the southeastern Bering Sea. In The eastern Bering

Sea shelf: Oceanography and resources, pp. 993-1012. Ed. by D. W. Hood and J. A. Calder. Univ. Washington Press, Seattle, 1339 pp.

Otto, R. K. 1981. Eastern Bering Sea crab fisheries. In The eastern Bering Sea shelf: Oceanography and resources, pp. 1037-1066. Ed. by D. W. Hood and J. A. Calder. Univ. Washington Press, Seattle, 1339 pp.

Otto, R. K. [1984]. Northwest and Alaska Fisheries Center, Nat. Mar. Fish. Serv., Kodiak, Unpubl. data and personal communication.

Overland, J. E., and Pease, C. H. 1982. Cyclone climatology of the Bering Sea and its relation to sea ice extent. Mon. Wea. Rev. 110:2015-2023.

Paluszkiewicz, T., and Niebauer, H. J. 1984. Satellite observations of circulation in the eastern Bering Sea. J. Geophys. Res. 89:3663-3678.

Paul, A. J. 1983. Light, temperature, nauplii concentrations, and prey capture by first feeding pollock larvae, Theragra chalcogramma. Mar. Ecol.-Prog. Ser. 13:175-179.

Paul A. J., Paul, J. M., Shoemaker, P. A., and Feder, H. M. 1979. Prey concentrations and feeding response in laboratory reared stage-one zoeae of king crab, snow crab, and pink shrimp. Trans. Am. Fish. Soc. 198:440-443.

Pearson, C. A., Mofjeld, H., and Tripp, R. B. 1981. Tides of the eastern Bearing Sea shelf. In The eastern Bering Sea shelf: Oceanography and resources, pp. 111-130. Ed. by D. W. Hood and J. A. Calder. Univ. Washington Press, Seattle, 1339 pp.

Pease, C. H. 1980. Eastern Bering Sea ice processes. Mon. Wea. Rev., 108:2015-2023.

Rogers, J. C. 1981. The north Pacific oscillation. J. Climatology 1:39-57.

Sambrotto, R. N., and Goering, J. J. 1983. Interannual variability of phytoplankton and zooplankton production on the southeast Bering Shelf. In From year to year: Interannual variability of the environment and fisheries of the Gulf of Alaska and the eastern Bering Sea, pp. 161-177. Ed. by W. S. Wooster. Washington Sea Grant Publ. No. WSG-WO 83-3, Seattle, 208 pp.

Sambrotto, R. N., and Goering, J. J. 1984. Relevancy of southeast Bering Sea oceanographic studies to fisheries and marine mammals. In Proc. Workshop on biological interactions among marine mammals and commercial fisheries in the southeastern Bering Sea, pp. 17-38. Alaska Sea Grant Rep. 84-1, Alaska Univ. 300 pp.

Schumacher, J. D., and Aagaard, K., Pease, C. H., and Tripp, R. B. 1983a. Effects of a shelf polynya on flow and water properties in the northern Bering Sea. J. Geophys. Res. 88:2723-2732.

Schumacher, J. D., and Incze, L. S. [1984]. Northwest Fisheries Center, Nat. Mar. Fish. Service, Seattle, Unpubl. data.

Schumacher, J. D., and Kinder, T. H. 1983. Low frequency current regimes over the southeastern Bering Sea Shelf. J. Phys. Oceanogr. 14:607-623.

Schumacher, J. D., Kinder, T. H., and Coachman, L. K. 1983b. Eastern Bering Sea. Rev. Geophys. Space Phys. 21:1149-1153.

Schumacher, J. D., Kinder, T. H., Pashinski, D. J., and Charnell, R. L. 1979. A structural front over the continental shelf of the eastern Bering Sea. J. Phys. Oceanogr. 9:79-87.

Schumacher, J. D., and Moen, P. 1983. Hydrography and circulation in Unimak Pass and over the shelf north of Alaska Peninsula. U.S. Dep. Commer., NOAA Tech. Memo. ERl/PMEL-47, 75 pp.

Schumacher, J. D., Pearson, C. A., and Overland, J. E. 1982. On exchange of water between the Gulf of Alaska and the Bering Sea through Unimak Pass. J. Geophys. Res. 87:5785-5795.

Schumacher, J. D., and Reed, R. K. 1983. Interannual variability in the abiotic environment of the Bering Sea and the Gulf of Alaska. In From year to year: Interannual variability of the environment and fisheries of the Gulf of Alaska and the eastern Bering Sea, pp. 111-133. Ed. by W. S. Wooster. Washington Sea Grant Publ. No. WSG-WO 83-3, Seattle, 208 pp.

Serobaba, I. I. 1968. Spawning of the Alaska pollock, Theragra chalcogramma (Pallas) in the northeastern Bering Sea. J. Ichthyol. (Engl. transl. Vopr. Ikhtiol.) i:789-798.

Shimada, A. M. [1984]. Northwest and Alaska Fisheries Center, Nat. Mar. Fish. Serv., Seattle, Unpubl. data.

Smith, G. B. 1981. The biology of walleye pollock. In The eastern Bering Sea shelf: Oceanography and resources, pp. 527-552. Ed. by D. W. Hood and J. A. Calder. Univ. Washington Press, Seattle, 1339 pp.

Smith, G. B., Allen, M. J., and Walters, G. E. 1984. Relationships between walleye pollock, other fish, and squid in the eastern Bering Sea. In Proc. Workshop on walleye pollock and its ecosystem in the eastern Bering Sea, pp. 161-191. Ed. by D. H. Ito. NOAA Tech. Memo. NMFS F/NWC-62, 292 pp.

Smith, S. L., and Vidal, J. 1984. Spatial and temporal effects of salinity, temperature and chlorophyll on the communities of zooplankton in the southeastern Bering Sea. J. Mar. Res. 42:221-257.

Smith, S. L., and Vidal, J. In press. Variations in
the distribution, abundance and development of
copepods in the southeastern Bering Sea in 1980
and 1981. Cont. Shelf. Res.
Somerton, D. A. 1981. Life history and population
dynamics of two species of Tanner crab, Chionoecetes bairdi and C. opilio, in the eastern Bering
Sea with implications for the management of the
commercial harvest. Dissertation, Univ. Washington, Seattle, 220 pp.
Straty, R. R. 1981. Trans-shelf movements of Pacific
salmon. In The eastern Bering Sea shelf: Ocean-
ography and resources, pp. 575-595. Ed. By D. W.
Hood and J. A. Calder. Univ. Washington Press,
Seattle, 1339 pp.
Takahashi, Y., and Yamaguchi, H. 1972. Stock of the
Alaska pollock in the eastern Bering Sea. Bull.
Jpn. Soc. Sci. Fish. 38:418-419.
Takeuchi, I. 1962. On the distribution of zoea larvae
of king crab, Paralithodes camtschatica, in the
southeastern Bering Sea in 1960. Bull. Hokkaido
Reg. Fish. Res. Lab. 24:163-170.
Vidal, J. 1980. Physioecology of zooplankton. I.
Effects of phytoplankton concentration, tempera-
ture and body size on growth rate of Calanus
pacificus and Pseudocalanus sp. Mar. Biol.
56:111-134.
Vidal, J., and Smith, S. L. In press. Biomass, growth
and development of populations of herbivorous zoo-
plankton in the southeastern Bering Sea during
spring. Deep Sea Res.
Wakabayashi, K. 1974. Studies on resources of the
yellowfin sole in the Bering Sea. I. Biological
characters. Fish. Agency Jap. (transl. available,
Nat. Mar. Fish Serv., Seattle), 77 pp.
Walsh, J. J. 1983. Death in the sea: Enigmatic
phytoplankton losses. Progr. Oceanogr. 12:1-86.
Weber, D. D. 1967. Growth of immature king crab
Paralithodes camtschatica (Tilesius). Bull. Int.
North. Pac. Fish. Comm. 21:21-53.
Weber, D. D. 1974. Observations of growth of
southeastern Bering Sea King crab, Paralithodes
camtschatica, from a tag-recovery study, 1955-65.
NOAA/Nat. Mar. Fish. Service Data Report No. 86.
Seattle, WA. 122 pp.
Wespestad, V. G. 1981. Distribution, migration and
status of Pacific herring. In The eastern Bering
Sea shelf: Oceanography and resources, pp. 509-
526. Ed. by D. W. Hood and J. A. Calder. Univ.
Washington Press, Seattle, 1339 pp.

8. Results of Recent Time-Series Observations for Monitoring Trends in Large Marine Ecosystems with a Focus on the North Sea

ABSTRACT

Examples of major changes in abundance and in biological parameters of some North Sea fish species are discussed, based on time-series observations derived from commercial catch data and research vessel surveys. It is concluded that the ongoing monitoring programs are capable of detecting long-term trends, but they have rarely added significantly to our understanding of the mechanisms behind the changes. There is a strong need for supplementing the extensive long-term monitoring programs by specific and detailed studies, which address testable hypotheses from the existing theory, in order to establish adequate management strategies for exploited large marine ecosystems.

INTRODUCTION

It will be understood that the amount of information available as time-series observations for the North Sea fish stocks is too unwieldy to be treated coherently in a short chapter. The symposium on the Changes in the North Sea Fish Stocks and their Causes held in 1975 (Hempel, 1978a) covers 449 pages, and since then extensive information has been added, much of which can only be extracted from "gray" literature. In this contribution I will highlight some examples of changes, draw attention to specific problems, and suggest how these might be approached by adjusting future sampling schemes.

Fisheries research is by its nature to a very large extent dependent on time-series observations. This is because fisheries do not lend themselves to rigid experimental testing of population dynamics theory. The only test may be found in the historic development of fish stocks in relation to their fisheries as measured in suitable population parameters.

145

Thus, from 1902 onward when the International Council for the Exploration of the Sea (ICES) was founded, detailed catch statistics by individual stock have been stressed as the primary data source for providing advice on fish stock management. The ICES Bulletins Statistiques still present the major source of information to stock assessment working groups in the Northeast Atlantic.

As fisheries science developed it was realized that effort data were virtually as important as total catches in analyzing fish stock dynamics. Within the last decade, extensive information on fleet statistics have been annually submitted to regional fisheries organizations. However, within ICES effort data have not been adequately standardized because the technology of fishing proceeded at a rate where it turned out to be almost impossible to standardize measures of effort satisfactorily, even over relatively short periods of time.

As the mathematical description of the dynamics of exploited fish stocks developed (Beverton and Holt, 1957; Ricker, 1958), the need for age structured catch data increased, and biological information on size and age distributions has been routinely collected in national sampling schemes. The introduction of the Virtual Population Analysis (VPA; Gulland, 1965), which allows the estimation of fishery mortalities when catches in numbers by age group are reported for all the major fisheries on a particular stock, provided an invaluable alternative for the more troublesome catch per unit of effort analysis. Although the VPA is still considered to be an excellent tool for analyzing "historic" developments in a fish stock, a drawback is the inherent uncertainty with regard to the "present" state of the stock. New models are being developed which combine VPA and CPUE data to make up for this deficiency (Pope and Shepherd, 1982; Gudmundsson et al., 1982).

The potential yield of a fish population is largely dependent on the average growth of the species with age and on the level of recruitment. When catches are sampled for age distribution, mean length at age is obtained as a by-product. However, whereas age structured catch data are routinely reported to working groups, growth data are generally not as accessible. This is regrettable, because extensive information on possible changes in growth rate over time must be contained in national data files. For studying possible changes in the level of recruitment, the spawning stock biomass or total egg production must be considered. Studies of maturity and fecundity require specific sampling schemes, and time-series information on changes in these parameters are rather limited.

The major management tool accepted within the ICES area is the Total Allowable Catch (TAC). Obviously, this requires a timely and accurate prediction of the potential catch rate at least one year ahead. Since in heavily exploited fish stocks the recruiting year class generally represents a significant component of the total catch, there is a strong need to have reliable estimates of abundance of the year class before it recruits to the fishery. For this purpose, large scale internationally coordinated young fish surveys have been carried out since 1965, and as the time series builds up, a wealth of data is becoming available. Information on species of lesser economic importance, which is being collected as a by-product, may ultimately result in a better understanding of the behavior of the whole system.

Apart from the general failure of the TAC management scheme due to severe enforcement problems, the effect on the scientific advice itself has been markedly adverse because TAC regulations have produced a strong incentive to falsify reported catches. Not only do the fishing industries seek uncontrolled ways to land fish, but the responsible national management agencies also sometimes appear to adjust reported catches to the TAC level rather than expose themselves to blame during international negotiations. The inherent scientific problem to cope with the deteriorating catch statistics has in some instances been solved by collecting a shadow reporting system on a confidential basis from a subsample of trustworthy skippers. As a consequence, the total international catch used by assessment working groups may nowadays deviate considerably from the official catch statistics reported. Still, worries about the deteriorating basis for stock assessment steadily increase and the need for additional fisheries independent data is becoming more urgent every year. Although stock size estimates from various surveys are becoming more regularly available, logistics have prevented establishing an adequate range of rigidly designed and annually repeated surveys as continuous input for stock assessment. The difficulties originate from the number of species, each of which may set specific demands on survey timing and design, from the large areas to be covered and from the specific and variable interests of the different countries.

A recent advance in stock assessment in the North Sea is related to the development of a simultaneous multispecies VPA (Helgason and Gislason, 1979; Pope, 1979; Sparre, 1980), which evolved from the highly detailed and complex ecosystem model developed by Andersen and Ursin (1977). Although as yet this approach has not been used for preparing management advice, its formulation emphasized the lack of suitable information on interspecific predation rates between exploited fish species. Following the identification of this information gap, a large scale internationally coordinated stomach sampling project was initiated in 1981. The results obtained so far (Daan, 1983) imply a major breakthrough in research, and the potential impact on management advice has been well identified. Although the estimated level of predation mortality among juvenile commercial fish species is strictly speaking not the result of time-series observations, a short review is included because of its possible impact on the range of future time series to be collected.

SOURCES OF INFORMATION

The catch statistics reported to Bulletins Statistiques have been used to split the total North Sea fish catch into 3 categories: pelagic, demersal, and industrial species. This split is somewhat artificial because part of the pelagic or demersal species may be exploited in the industrial fisheries, whereas the industrial species could be assigned to either the pelagic or the demersal group. The criterion used has been that industrial species only comprise those species, which are too small for human consumption and which are almost exclusively exploited by the fisheries for fish meal (Norway pout, sprat, sand eels). The pelagic component includes those pelagic, schooling fish which are at least partly exploited for human consumption (herring, mackerel, sardines, horse mackerel). The remainder, which may be caught in the industrial fisheries as a bycatch, are taken together as demersal fish (gadoids, flatfish, etc.). It was in this case impossible to make adjustments for possible deviations from the official statistics as used by working groups, because the availability of "unreported" landings varies extensively between species.

The information on catches, spawning stock biomasses, fishing mortalities, and recruitment presented in the three examples (herring, cod, and Norway pout) represent an update of Daan (1980) based on data presented by various recent assessment working groups. It should be noted that because the various North Sea fish stocks sometimes represent two or more substocks, which spawn at different times and in different areas, and are also exploited differently, the situation is in fact rather more complicated than presented in this overview.

Growth data have been selected from the available literature. A problem with the interpretation is that they refer generally to results obtained from specific fisheries and therefore they may not be representative for the North Sea stocks as a whole.

In 1964 young herring surveys were initiated by the Netherlands. Initially, these surveys were restricted to the area of high abundance of juvenile herring, but as more countries joined the exercise the coverage was gradually extended to include the entire North Sea. Thus it was possible to address various other species, particularly cod, haddock, whiting, and Norway pout, and the survey developed into a general International Young Fish Survey (IYFS). At present between 300 and 400 trawl hauls are made during a synoptic survey in February in which nine countries and even more vessels participate. There have been considerable logistic problems in coordinating these surveys in terms of standardization of gear and sampling procedures as well as in exchange of data. Therefore, as a time series there are problems due to the repetitive changes as the survey developed. Detailed evaluations have been reported regularly (e.g., Anon., 1981, 1983).

Following a recommendation of the ICES ad hoc Working Group on Multispecies Assessment Model Testing, a large-scale stomach sampling project was carried out in 1981. The five species among the exploited fish species complex, which were considered to represent the potentially most important fish predators, were sampled quarterly during surveys covering the entire North Sea. Over 40,000 stomachs were analyzed to estimate the average stomach content of the predators by age group, taking into account the spatial distribution of the predators. For the exploited fish species in the diet, the average number present in the stomach were also estimated by age group. A review of the logistics of the project as well as details on sampling intensity and preliminary results are summarized by Daan (1983).

TIME-SERIES OBSERVATIONS

Catch Trends by Categories

The total North Sea fish catch (Figure 8.1(a)) increased from 1 million tons at the beginning of the century to 2 million tons in 1956, with marked interruptions caused by the two World Wars. In the late 1950s catches declined to less than 1.5 million tons, but a pronounced increase followed in the early 1960s. For a decade the catch fluctuated rather steadily around 3 million tons, but in recent years the catch has fallen back to 2.5 million.

When split into the three categories, pelagic, demersal, and industrial species, it becomes clear that

tonnes x 10^{-6}

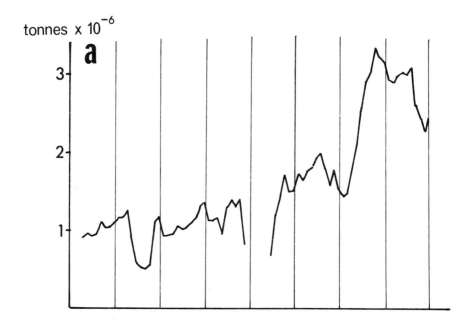

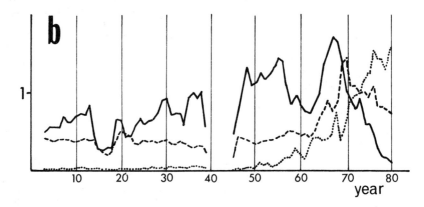

year

Figure 8.1. Trend in total catch (a) and for three categories of fish species separately (b). Pelagic species (solid line); demersal species (dashed line); and industrial species (dotted line).

until the late 1950s the catch of demersal species remained remarkably stable, and the annual fluctuations in the total catch just reflect the variations in pelagic catch (Figure 8.1(b)). The change in total catch in the early 1960s, however, coincided with a marked change in catch composition. Initially an increase was observed in all three categories, but the increase in pelagic yield was of a very short duration and a sharp decline followed. By 1980 the pelagic species contributed only a fraction of the long-term average.

The demersal species reached a maximum some years later than the pelagic species and the decline did not bring them nearly to the lower levels of the herring. In fact, in recent years the catch is still higher than ever recorded before 1965. Finally, the industrial species appear also to have reached a ceiling by the end of the 1970s and their catch has, as yet, not regressed.

Although the catch composition has undergone major changes after a long period of apparent stability, it is impossible to judge without any additional information whether these changes reflect a real change in the species composition of the ecosystem or just trends in the directed fishing effort. Undoubtedly, over this century fleets have undergone marked changes. For instance, the development of more powerful engines allowed the introduction of the herring trawl fishery after the Second World War. The passive drift net fishery disappeared, and the active search for herring schools by means of hydroacoustic equipment became a major component of the fishing effort. In the 1960s with the introduction of the purse seine, the catch potential of a vessel became almost solely dependent on the ability to locate schools and on the loading capacity of the vessel. This also changed the nature of the fishery, because the huge catch rates of the purse seine fleet did not allow processing of the fish caught for human consumption and a considerable part of the total herring and mackerel catches were turned into fish meal.

In the demersal fisheries, changes in efficiency were less pronounced. Whereas the Danish seine survived in traditional fisheries, long lining disappeared and steam trawling was gradually replaced by motor trawling. Some new techniques such as pair trawling and gill netting were introduced and in the flatfish sector the very efficient beam trawl became an important gear.

Particularly noteworthy is the development of the industrial fishery in the 1960s and 1970s, which includes both purse seiners and relatively small trawlers employing small meshed nets.

In conclusion, there have been quite a number of qualitative changes in effort, which undoubtedly have had their effect on the changes in total catch composition. To what extent the species composition in the sea has changed remains unclear. In an effort to identify the extent of changes in species composition, I have examined the population dynamics of three North Sea fish species--herring, cod, and Norway pout--as examples of the need for more detailed treatment of population measurement parameters.

EXAMPLES

Four main parameters by which the development of an exploited fish population can be described are the annual catch (a function of population size and fishing effort), the spawning stock biomass (which provides the basis for replacement within the stock), the fishing mortality coefficient (a direct measure of the fishing effort to which the population is subjected), and the annual recruitment. Estimates of these four parameters as time-series should make it possible to judge the changes that have taken place within fish stocks.

Herring

The herring has in the past provided the major component of the North Sea pelagic yield and the total collapse of the pelagic fisheries is well illustrated by this species. The catch (Figure 8.2(a)) fluctuated between 600 and 800 thousand tons during most of the century until in the mid 1960s the catch peaked over 1 million tons. Directly after this, a dramatic decline followed, and in 1977 the fishery was closed. Only minor catches were maintained in industrial fisheries until 1982, when the fishery was opened again under a TAC management regime.

Estimates of the spawning stock biomass (Figure 8.2(b)) reveal that the stock has been declining from the second World War onwards with two revivals of short duration in 1959 and 1964. Apparently, the biomass had been decimated already by 1969 even though the total catch remained high for another few years and the decision to close the fishery in 1977 was taken when there was hardly any herring left. Rebuilding of the stock was initially slow and it took 5 years (1982) before the stock had increased to a level, where a small quota could be allowed.

The estimated fishery mortality (Figure 8.2(c)) indicates a gradual increase in exploitation from 1947 onwards. The reduction in biomass directly after the second World War, however, is probably more closely related to the fact that the stock had been able to

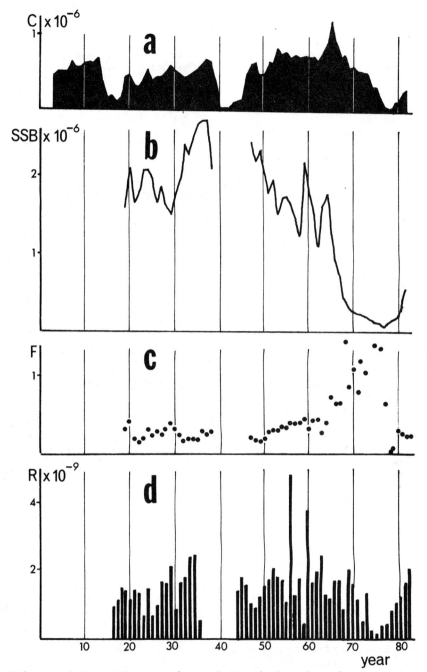

Figure 8.2. Time series of North Sea herring population parameters: (a) catch; (b) spawning stock biomass; (c) fishery mortality coefficient; (d) recruitment.

build up in the absence of fishing, rather than to increased exploitation. In the mid-1960s the introduction of the purse seine led to considerably increased fishery mortalities and these must be held responsible for the further decline in spawning stock biomass. Because exploitation rate remained high until the fishery was closed in 1977, the catch level did not nearly reflect the reduction of the stock that had taken place over several years.

The time series of recruitment (Figure 8.2(d)) indicates considerable fluctuations. The year classes 1956 and 1960 were particularly outstanding and their affect on the biomass some three years later can be easily identified. After the average year class of 1973 a series of four years followed, in which recruitment remained lower than ever recorded before and undoubtedly this must have been related to the very low stock sizes during this period. When the first extremely poor year class appeared, the herring stock was assessed to be in a state of recruitment overfishing, and the associated imminent danger for the stock has been the main argument for closing the fishery.

The stock recruitment plot (Figure 8.3) indicates the seriousness of the situation. Still, over a very wide range of stock biomasses recruitment has been virtually constant, although the suggestion could be made that at high stock levels variability of recruitment is increased.

Altogether the North Sea herring appears to provide a coherent picture of interaction between catch, biomass, exploitation, and recruitment. However, it must be stressed that when the various substocks are considered separately, the situation becomes rather more complicated and does not seem nearly as simple (Burd, 1978). The closing of the fishery prevented monitoring the recovery of the stock by employing fishery-independent methods.

Cod

The development in the cod stock (Figure 8.4) as an example of a demersal species has been entirely different. Again the catch remained constant over most of the century (Figure 4(a)), but in the mid-1960s suddenly doubled and remained at that level. Also the amplitude of the annual fluctuations appears to have increased. The increase in catch is associated with a comparable increase in spawning stock biomass, but in recent years biomass has fallen back to the original level (Figure 8.4(b)).

The estimated fishery mortalities (Figure 8.4(c)) do not indicate any significant increase at the time when catch and biomass increased, but the increased

155

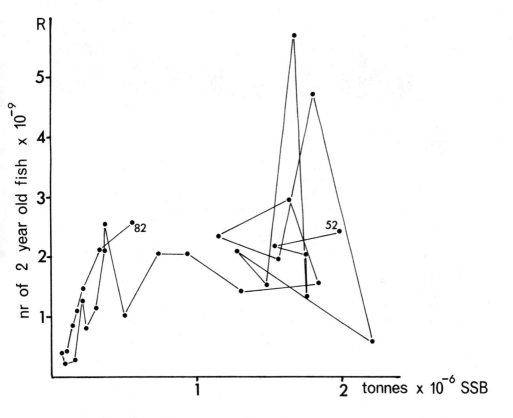

Figure 8.3.   Stock-recruitment plot of North Sea herring.

156

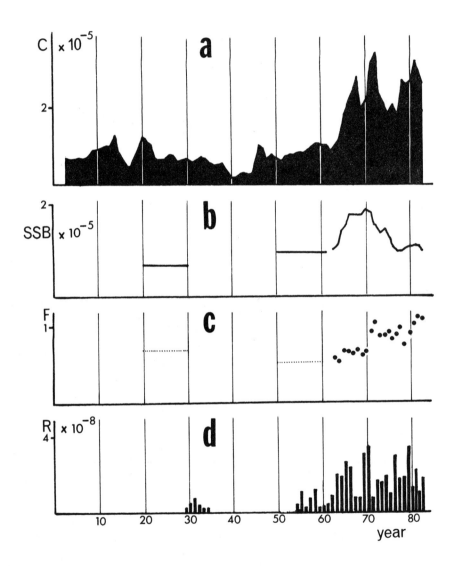

Figure 8.4. Time series of North Sea cod population parameters: (a) catch; (b) spawning stock biomass; (c) fishery mortality coefficient; (d) recruitment.

exploitation observed in the 1970s may have been responsible for the decrease in biomass. The total catch does not seem to have been affected.

Data on the pre-1960 situation in the cod stock are not nearly as good as for herring, but from the scattered data available from various sources the increased level of recruitment has been well established (Figure 8.4(d)). Even the poor year classes in recent years appear to be more numerous than before. Thus, the rapid increase in biomass and catch does not seem to be related to a marked change in the fishery for cod and the cause must be found in the sequence of a large number of strong year classes during the last two decades. Apparently, the carrying capacity of the system for the juvenile life stages has changed, but how and why this happened remains unsolved.

A plot of recruitment versus spawning stock biomass (Figure 8.5) only provides a scatter of points all over without indicating any particular pattern.

Norway Pout

The industrial fishery for Norway pout developed only recently and the observed increase in catch (Figure 8.6(a)) is undoubtedly related to the expansion of the fishery. As one would expect of a short-lived species, the catches as well as the spawning stock biomass (Figure 8.6(b)) fluctuate considerably from year-to-year. In addition, the industrial fleet has some choice of concentrating on other species, whichever is most abundant, and this adds another element of variation to the catches.

A problem with species characterized by high natural mortalities is that the traditional assessment methods (VPA) do not work very satisfactorily and the population parameters obtained cannot be considered very reliable. As yet, however, there is no obvious alternative. Moreover, the range of years over which biomass estimates and fishery mortalities (Figure 8.6(c)) are available is limited, so that long-term trends cannot be established. The year-class strength indices (Figure 8.6(d)) have been taken from the IYFS and also refer to the most recent period only.

Some clue, however, can be found in the catch rates of Norway pout in Scottish research vessel surveys, which have been carried out with an interruption between 1920 and 1968 (Richards et al., 1978). These data, indicated by the interrupted line in Figure 8.6(b) suggest that the stock has gradually increased after the Second World War, approximately a decade before the fishery developed and also a decade before the large changes observed in the herring and cod stocks.

158

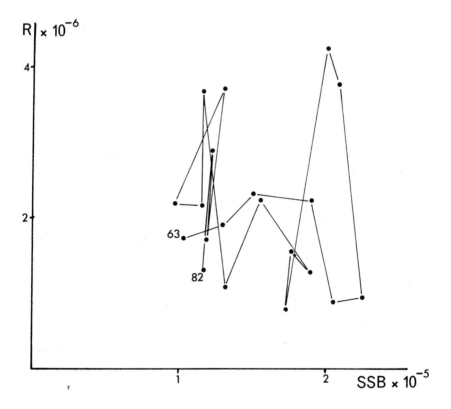

Figure 8.5. Stock-recruitment plot of North Sea cod.

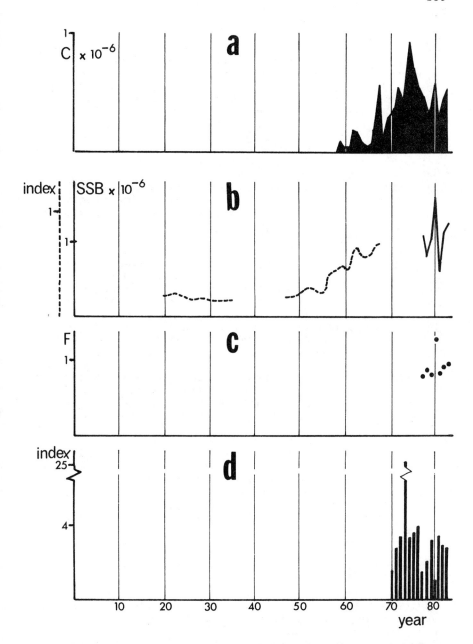

Figure 8.6. Time series of North Sea Norway pout population parameters: (a) catch; (b) spawning stock biomass; (c) fishery mortality coefficient; (d) recruitment.

Survey Data

In the three examples given, the main body of the population parameters presented have been deduced from the application of estimation models to age structured catch data (VPA). These methods involve a large and variable amount of uncertainty, due both to imperfect data and to arbitrary assumptions. This uncertainty cannot be properly assessed because in a multicountry situation with considerable variations in sampling schemes, it is virtually impossible to calculate error bounds around estimated parameter values. Secondly, the VPA methods suffer inherently from the larger number of unknowns than there are basic equations, and in order to solve for fishery mortalities, natural mortality has generally to be assumed constant at a rather arbitrary value.

As a check on the general validity of VPA methods, the annual estimates of recruitment derived from these can be correlated with indices of abundance from surveys (Figure 8.7). Although in all four species presented here there is a significant correlation between the two independent estimates, it is quite clear, that in all cases the variance around the estimated geometric regression lines is considerable. An abundant year class in the survey may turn up only as an average one in the VPA and vice versa.

Recently, a comparison between various independent survey indices for individual year classes in the North Sea indicated that, in fact, such indices are generally rather better correlated than any of those with VPA recruitment figures (Anon., 1983). This suggests that surveys do provide a more reliable index of the number of fish present in the sea than the VPA, and to improve the correlations and our confidence in assessment, it would seem more profitable to try to get rid of some of the stringent assumptions underlying the VPA than to intensify the surveys. Particularly, the assumption of a constant natural mortality over all age groups and years could introduce a severe bias in the estimates of recruitment and should probably be amended.

Trends in Growth

As stated earlier, the catch of a fish species reflects the interaction of stock size and fishing effort, but more precisely the yield will also be a function of the rate of growth of the individuals. Since any density-dependent changes in growth rate as a response to changes in exploitation could have considerable effect on the yield potential, it would seem worthwhile to investigate how strong the evidence for

161

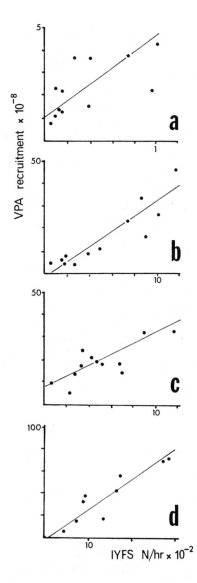

Figure 8.7. Correlations between VPA estimates of recruitment vs. IYFS indices of abundance: (a) cod (N = 12; r = 0.58; p <0.05); (b) haddock (N = 12; r = 0.92; p <0.005); (c) whiting (N = 12; r = 0.77; p <0.005); (d) herring (N = 9; r = 0.91; p <0.005).

density-dependent growth is by discussing two examples
of observed changes.

Figure 8.8(a) depicts the trend in mean length at
3 yr old for a substock (Buchan) of the North Sea
herring according to Hubold (1978). Since the origin
of these variations could be traced back to the length
at 1 yr old by back calculation from otolith rings, in
Figure 8.8(b) the estimated lengths at 1 yr old are
plotted by the author against stock size in numbers of
1-, 2-, and 3-yr-old herring. As one might expect,
there is a negative correlation since there is a
positive trend in mean length over time and a negative
trend in biomass over the same time period (cf Figure
8.2(b)). In dealing with parameters showing time
trends there may be pitfalls in concluding a causal
relationship. This is particularly true in this case
with reference to density dependent growth, because the
change in mean length (Figure 8.8(a)) appears to
develop rather consistently from year-to-year, whereas
the residuals of the correlation with abundance appear
to be very much higher (Figure 8.8(b)). More likely, a
steady trend in growth rate could be caused by a
genetic change in the population: under the stress of
heavy exploitation fast-growing individuals reaching
maturity at an early age might have a competitive
advantage over the slow-growing ones, because the
latter are vulnerable to the fishery for a longer
period before they reach maturity.

In Figure 8.9(a) mean length at 3 yr old for North
Sea cod are plotted against time, based upon two
sources of information (Macer, 1983, and Van Alphen,
1984). The English fishery for cod extends more nor-
therly than the Dutch fishery and the difference in
mean length can be explained by the fact that these two
fisheries exploit different parts of the cod stock.
Broadly speaking, however, both sets exhibit similar
trends in the variations from year-to-year. Macer
(1983) correlated mean length with the estimated
average total cod biomass in the year, when the year
class was 1 yr old (Figure 8.9(b)). Van Alphen (1984),
on the other hand, chose to correlate mean length with
the number of the same year class at the same age
estimated from VPA. In both cases significant negative
correlations were obtained, which have been interpreted
as evidence of density dependent growth. There are
some problems of a statistical nature because both
authors have tried several measures, which spoil the
value of the correlation coefficient as a measure of
significance. However, more importantly, when comparing these results, it is quite clear that they in
fact point to rather different density dependent
mechanisms. Macer's (1983) data imply that the
total stock biomass determines the growth rate of

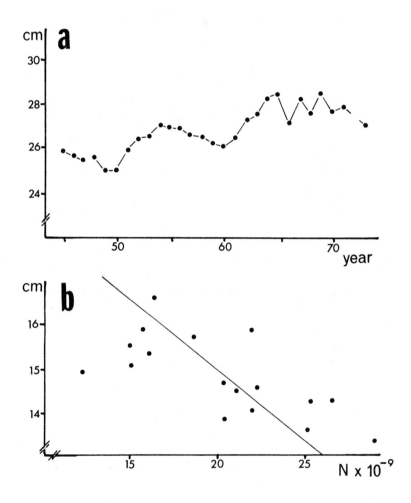

Figure 8.8. (a) Mean length of Buchan herring at 3 yr old. (b) Relation between length at age 1 and stock size of 1-, 2-, and 3-yr-old herring (r = 0.66; p <0.05). Redrawn after Hubold (1978).

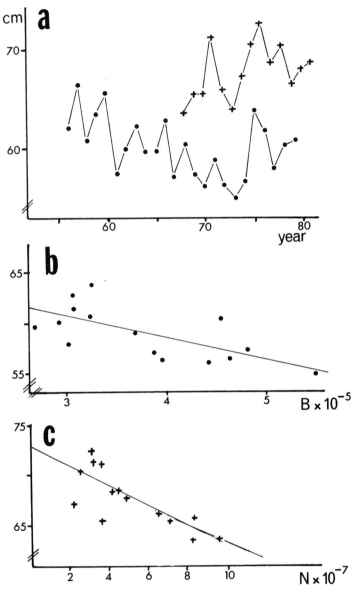

Figure 8.9. Mean length of cod at 3 yr old in the English (o; redrawn after Macer, 1983) and Dutch (+; redrawn from Van Alphen, 1984) fisheries. (a) Plotted as a time series. (b) Relation between English length at age and estimated average total stock biomass in the year, when the year class was 1 yr old ($r = -0.69$; $p < 0.01$). (c) Relation between Dutch length at age and estimated number of fish at 3 yr old from VPA ($r = -0.77$; $p < 0.005$).

1-yr-old cod and that the length increment during this year of life would set the handicap for the rest of its life. However, it is difficult to visualize how the small cod could compete for food with the older age groups because the food spectrum changes entirely with age (Daan, 1973, 1983). Van Alphen (1984) found that the number of cod at age 3 determines size at age 3, which would point to competition within an age group as the density dependent mechanism. Since both size and density at age 3 are probably correlated with size and density at earlier ages, the actual competition might in this case also take place earlier in life, but this could not be further analyzed on the basis of data from the commercial fisheries because it is only at 3 yr old that cod are fully recruited to the fishery.

To investigate the possible effect of density on size during the first two years of life, information from the International Young Fish Surveys (IYFS) has been plotted in Figure 8.10, but neither for 1- nor 2-yr-old cod is the correlation significant. Evidently, the problem of density dependent growth has not been solved satisfactorily for cod. In this respect, it should be noted that in analyzing length-at-age data from the fishery one might not obtain a reliable measure of growth in the sea because the fishery itself reacts to variations in density. Strong year classes generally attract the fishery, and at increased levels of fishery mortality the mean size of the fish caught will be smaller, irrespective of growth rate.

Also, fisheries may shift from one area to another depending on the density. Since considerable spatial differences in size at age exist according to survey results (Figure 8.11), annual variations in distribution of effort may easily result in variations in mean size at age in the catch, which are related to density through the fishery but not to density dependent growth.

QUANTITATIVE INTERACTIONS

The marked changes in abundance of the species discussed here, which present only a subset of changes observed (Hempel, 1978a, 1978b), pose the question if, apart from the impact of the fisheries on the individual stocks, these are in some way causally related through interspecific interactions. In respect to fish stock management this would be of particular interest because measures that would be profitable for one fishery might have adverse effects on other species. There appear to be two main lines along which interactions might operate: competition for a common food

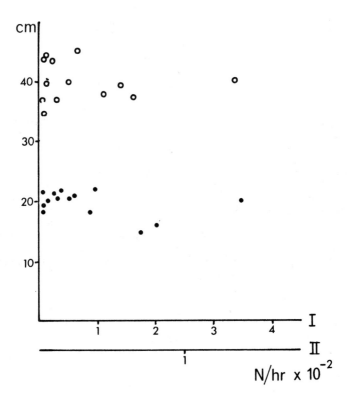

Figure 8.10. Relation between mean size of cod at 1 (●) and 2 yr old (o) as estimated from IYFS and the index of abundance during these surveys (1969-1982).

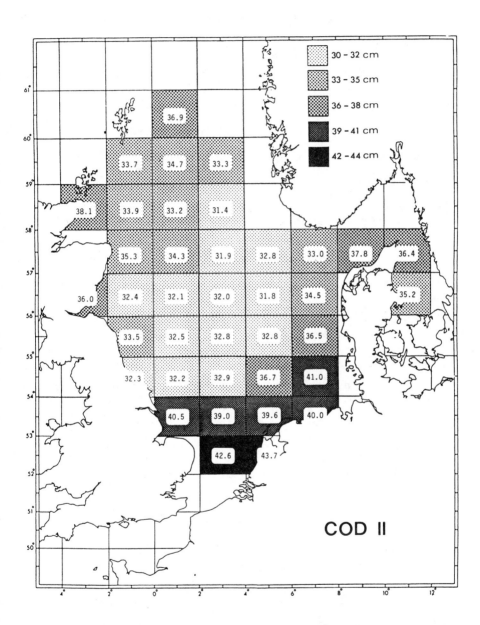

Figure 8.11. Spatial differences in mean length of 2-
yr-old cod based on IYFS data (long-term averages 1970–
1980).

resource and interspecific predation. However, as concluded by Sissenwine (in press) in a comprehensive review of these aspects in relation to recruitment, recent evidence appears to point to predation as being a more important factor than competition.

Predation may affect population dynamics of a fish stock at two levels. Firstly, the natural mortality in the fishable stocks may vary between years and between age groups depending on the abundance of predator stocks, and allowance for such variation should improve the assessment of the stock. This problem is relatively easily addressed by sampling predator populations, quantifying predation, and taking into account the age structure of the prey. Secondly, predation on eggs and larvae may be associated with the carrying capacity of the ecosystem for a particular species and thus influence recruitment. In contrast with predation on the fishable biomass, where the majority of the main predators may be represented by exploited fish species and can thus be easily studied, the direct interactions between exploited fish species in the egg and larval phase are more likely to represent only a relatively small component of the total predation. Also, this predation may be limited in time and space scales and therefore be more difficult to address. Some recent observations, however, will be discussed.

Stomach Sampling Project

The main aim was to provide estimates of the average stomach contents by age group of five predatory fish species. Prey were recorded by size classes, so that they could be classed in age groups by applying appropriate age size keys. This information, coupled with data on digestion rates to estimate food consumption, was required to tune the preference matrix related to the inter- and intraspecific predation by age group in the multispecies version of the VPA.

In Table 8.1 the 1981 estimated predation rates by the various predators combined are compared with the estimated stock sizes by age group on the basis of traditional assumptions of constant natural mortality at a fixed value. This exercise results, in a number of instances, in severe discrepancies. For instance, for 1-yr-old cod, haddock, whiting, and herring, apparently the predators have removed more fish than the working groups have estimated to have been in the sea. The only conclusion can be that the coefficients of natural mortality used are in general much too low for the younger age groups. More importantly, however, it appears highly unlikely that under the changes in abundance observed in, for instance the cod stock, such

TABLE 8.1.

Estimated numbers of various exploited fish species by age group consumed by cod, whiting, and saithe. Predation (P) in 1981 shown in comparison with estimated numbers in the sea at the beginning of 1981 (N) from VPA (F = fraction consumed, revised from Daan, 1983). (N and P in '000,000 fish).

Age	Cod			Haddock			Whiting		
	N	P	F	N	P	F	N	P	F
0	?	6907	?	2278	24200	10.62	1604	8537	5.32
1	131	163	1.24	341	2112	6.19	498	949	1.91
2	313	18	.06	1018	235	.23	893	274	.31
3+	76	2	.03	295	11	.04	662	69	.10

Age	Norway Pout			Herring			Sprat		
	N	P	F	N	P	F	N	P	F
0	232505	45682	.20	12414	7625	.61	3781	2209	.58
1	21971	15434	.70	2347	3742	1.59	21458	9128	.43
2	14634	5018	.34	1496	69	.05	5176	4276	.83
3+	383	208	.54	1147	127	.11	244	287	1.18

predation mortalities would be constant from year-to-year.

The real impact of predation on the population dynamics of the various fish species cannot be assessed at this stage. But the application of multispecies VPA methods which take the interspecific predation into account in the near future will undoubtedly change our ideas significantly of what has happened in the North Sea fish stocks.

Interactions in the Egg Phase

During three spawning seasons of plaice and cod (1980, 1982, and 1983), stomach samples of herring have been collected with a view of estimating the number of eggs consumed during the spawning season in comparison with the estimated numbers of eggs produced (Daan et al., 1984). Again sampling was carried out over the total area of distribution of the herring in order to take into account the spatial distribution of the predator as well as of the rate of egg predation. The results are summarized in Table 8.2. There are still considerable uncertainties in the procedure of calculations because the time it takes to digest fish eggs in herring stomachs could well be rather faster than the 12 h assumed here. For the time being, however, it would seem that a fraction of 1% for plaice and 0.1% for cod would not nearly be enough for the herring stock to have a large impact on the recruitment of these species, even when the herring stock in those years was still a fraction of the situation before the 1960s.

DISCUSSION

The picture I have presented here of the changes in the North Sea fish stocks is far from complete. However, it is quite clear that the ongoing monitoring programs in terms of catch statistics and surveys have been insufficient to analyze the mechanisms underlying the interactions between the various fish species even if time series of some of the important population parameters could be established. In other words, it remains uncertain if the increase in demersal and industrial stocks are causally linked to the collapse of the pelagic stocks, or what will happen when and if the herring stocks recover.

Because of the deterioration of catch statistics under the present TAC management system, it is expected that an increasing proportion of the research effort will be spent on the collection of fishery-independent data, in particular from surveys. It seems absolutely critical that in planning those surveys more emphasis

TABLE 8.2.

Estimated egg production (E) of North Sea cod and plaice in comparison with the estimated egg predation (P) by the North Sea herring stock for 1980, 1982, and 1983. (F = fraction consumed; from Daan et al., 1984). (E and P in '000,000,000).

Year	Species	E	P	F
1980	Cod	40	.02	.0005
	Plaice	33	.25	.0076
1982	Cod	39	.04	.001
	Plaice	36	.50	.014
1983	Cod	32	.06	.002
	Plaice	42	.81	.019

be placed on collecting quantitative information related to the likely causal interactions between state variables. Progress in fisheries science appears to be less dependent on the addition of just another year's stock sizes than on the careful collection of specific data, which allow testing of existing theory. As such the stomach sampling project may serve as an example of how routine surveys may be used at relatively low extra costs to quantify predation mortality. But there may be other aspects such as growth and maturation as well, which are better studied on the basis of survey material rather than from sampling market statistics.

Undoubtedly, any serious attempt for adequate future management of LMEs will depend heavily on the willingness to support extensive and appropriate monitoring programs over long time spans. This is not just a scientific requirement, but more importantly, management agencies should demand that such information be collected in order to allow evaluation of their strategies. Whereas it seems unlikely that fishery scientists will be able to predict any large scale interspecific consequences of major management measures in the near future, a sound monitoring program could facilitate an empirical approach in managing ecosystems.

REFERENCES

Andersen, K. P., and Ursin, E. 1977. A multispecies extention to the Beverton and Holt theory of fishing with accounts of phosphorus circulation and primary production. Meddr Danm. Fisk.- og Havunder., N.S., 7:319-435.

Anonymous. 1981. Report of the Joint Meeting of the International Young Herring Survey Working Group and the International Gadoid Survey Working Group. ICES C.M. 1981/H:10.

Anonymous. 1983. Report of the International Gadoid Survey Working Group. ICES C.M. 1983/G:62.

Beverton, R. J. H., and Holt, S. J. 1957. On the dynamics of exploited fish populations. Fishery Invest., Lond. (2) 19:533 pp.

Burd, A. G. 1978. Long-term changes in North Sea herring stocks. Rapp. P.-v. Réun. Cons. int. Explor. Mer 172:137-153.

Daan, N. 1973. A quantitative analysis of the food intake of North Sea cod, Gadus morhua. Neth. J. Sea Res. 8(1):27-48.

Daan, N. 1980. A review of replacement of depleted stocks by other species and the mechanisms underlying such replacement. Rapp. P.-v. Réun. cons. int. Explor. Mer 177:405-421.

173

Daan, N. 1983. The ICES Stomach Sampling Project in 1981: aims outline and some results. NAFO SCR Doc. 83/IX/93.

Daan, N., Rijnsdorp, A. D., and Overbeeke, G. R. 1984. Predation of plaice and cod eggs by the North Sea herring stock. Contribution to ELH Symposium, Vancouver.

Gudmundsson, G., Helgason, T., and Schopka, S. A. 1982. Statistical estimation of fishing effort and mortality by gear and season for the Icelandic cod fishery in the period 1972-79. ICES C.M. 1982/G:29.

Gulland, J. A. 1965. Estimation of mortality rates. Annex to Arctic Fisheries Working Group report. ICES C.M. 1965/Doc. no 3:9 pp.

Helgason, T., and Gislason, H. 1979. VPA-analysis with species interaction due to predation. ICES C.M. 1979/G:52.

Hempel, G. 1978b. North sea fisheries and fish stocks--a review of recent changes. Rapp. P.-v. Réun. Cons. int. Explor. Mer 173:145-167.

Hempel, G., (ed.). 1978a. North Sea fish stocks - Recent changes and their causes. Rapp. P.-v. Réun. Cons. int. Explor. Mer 172:449 pp.

Hubold, G. 1978. Variations in growth rate and maturity of herring in the northern North Sea in the years 1955-1973. Rapp. P.-v. Réun. Cons. int. Explor. Mer 172:154-163.

Macer, C. T. 1983. Changes in growth of North Sea cod and their effect on yield assessments. ICES C.M. 1983/G:8.

Pope, J. G. 1979. A modified Cohort Analysis in which constant natural mortality is replaced by estimates of predation levels. ICES C.M. 1979/H:16.

Pope, J. G. and Shepherd, J. G. 1982. A simple method for the consistent interpretation of catch-at-age data. J. Cons. Cons. int. Explor. Mer 40:176-184.

Richards, J., Armstrong, D. W., Hislop, J. R. G., Jermyn, A. S., and Nicholson, M. D. 1978. Trends in Scottish research-vessel catches of various fish species in the North Sea, 1922-1971. Rapp. P.-v. Réun. Cons. int. Explor. Mer 172:211-224.

Ricker. 1958. Handbook of computations for biological statistics of fish populations. Bull. Fish. Res. Board Can. 119:300 pp.

Sissenwine, M. (In press). Why do fish populations vary? In Exploitation of marine communities. Dahlem Workshop Reports.

Sparre, P. 1980. A goal function of fisheries (Legion analysis). ICES C.M. 1980/G:40.

Van Alphen, J. 1984. Growth and density of cod in the southern North Sea. Netherlands Institute for Fishery Investigations Intern. Rep. ZE 84-04 (mimeo; in Dutch).

9. Comparison of Continuous Measurements and Point Sampling Strategies for Measuring Changes in Large Marine Ecosystems

ABSTRACT

An undulating Batfish vehicle towed in a cyclic pattern while sampling conductivity, temperature, depth, in situ chlorophyll a, and copepods in the upper surface layers (0 to 100 m) has been used to obtain continuous series of profiles separated by ≈1 km or less horizontally in various marine ecosystems. These data have enabled statistical comparison of the vertical and horizontal variability of biological measurements which considered in general terms as biomass or more specifically as integrated chlorophyll ($mg \cdot m^{-2}$), copepods ($\# \ m^{-2}$), and primary production ($mg \ C \cdot m^{-2} \ d^{-1}$), the latter being generated from a numerical chlorophyll/light model. In the case of horizontal variability, single profiles are compared (taken here as point samples) against the profile series. Data were analyzed from various marine ecosystems consisting of the Eastern Canadian Arctic, Scotian Shelf, Peruvian Shelf, and Eastern Tropical Pacific. These geographical regions provide a wide range of biological growth periods from ≈1-2 months in the Canadian Arctic to year-round in the Tropical Pacific.

INTRODUCTION

Since Hardy (1956) reported large inter-station variability of ichthyoplankton samples in the southern North Sea, planktologists involved in biological surveys have been concerned with the statistical reliability of their plankton estimates from bottle samples and net hauls. Statistical treatment of the variability of these replicate samples have been since applied vigorously (Barnes and Marshall, 1951; Hasle, 1954; Platt, 1972). Characteristic length scales of 1 to 10 km for phytoplankton patches have been identified (Kierstead and Slobodkin, 1953; Denman and Platt, 1975; Powell et al., 1975; Fasham and Pugh, 1976; Steele and Henderson, 1977; Lekan and Wilson, 1978) while observations of zooplankton variability suggest patches exist typically on scales of several km (Wiebe et al., 1976) and 10 to 100 km (Steele, 1976). Unfortunately researchers cannot afford the time or luxury of stations 1 km apart while surveying a large area; yet by selecting stations 10 to 20 km apart, a survey may be undersampled.

This chapter examines the horizontal and vertical variability of three biological parameters, chlorophyll a, carbon production, and zooplankton (concentration per m^{-2}) in a variety of marine ecosystems. These data were obtained from the "Batfish," a towed undulating vehicle, which sampled sequential profiles spaced ≈ 1 km apart along transects 50 km in length, with a vertical resolution of ≈ 0.5 m. These data allowed comparison of errors in biological estimates made from "point samples," that is, comparisons between a single vertical profile vs. 50 profiles, for example, spaced ≈ 1 km apart if we consider horizontal scales.

Variability in biological estimates was compared between several geographical areas which afforded a wide range of physical control over biological growth. These areas were: Peruvian coastal waters, Eastern Tropical Pacific, Eastern Canadian Arctic (Lancaster Sound and Baffin Bay), and Scotian Shelf waters on the eastern Canadian coast. The Peruvian coast supports high productivity as a result of coastal upwelling and displays high variability (Herman, 1982; Herman, 1984) as a result of strong advection and vertical mixing. The Eastern Tropical Pacific represents a year-round near equilibrium of biological growth without large seasonal changes. The Scotian Shelf supports medium level of biological production which is seasonally variable. In contrast, the Eastern Canadian Arctic supports low production during a short summer growth period of ≈ 1-2 months.

SAMPLING AREA AND METHODS

The geographical locations (Figure 9.1) were in the Eastern Canadian Arctic (Baffin Bay and Lancaster Sound), Eastern Tropical Pacific, the Peruvian Coast, and the Scotian Shelf south of Nova Scotia. In each case, Batfish transects averaging 30 to 50 km in length were obtained with profile separations of about 0.7 km and depths ranging from near surface (3 m) to 100 m. Sampling was accomplished by towing a cycling Batfish vehicle in a sawtooth pattern while equipped with the following sensors: (1) a digital CTD unit (Model 8700 series, Guideline Instruments, Smith Falls, Canada); (2) a modified in situ Variosens fluorometer (Impulsphysik GmbH, Hamburg, West Germany); and (3) an

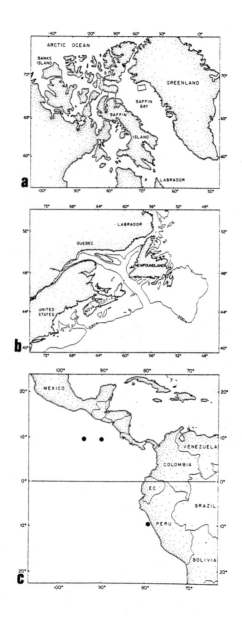

Figure 9.1.   Sampling areas used in the analysis of
Batfish data: (a) the Eastern Canadian Arctic (Baffin
Bay and Lancaster Sound), (b) the Scotian Shelf south
of Nova Scotia, and (c) the Eastern Tropical Pacific
and Peruvian Coast.

in situ electronic zooplankton counter (prototype). The system components included: (1) the Batfish vehicle (Dessureault, 1976), (2) in situ chlorophyll detection (Herman and Denman, 1976), and (3) in situ electronic zooplankton counter (Herman and Dauphinee, 1980).

The in situ electronic zooplankton counter was used to count and size copepods in volume ranges of 0.1 to 100 mm^3 which permitted discrimination of the dominant (3-4 species) copepod taxa present in the upper 100 m. The copepod identification procedure has been outlined in Herman and Mitchell (1981). In this chapter, we separate copepods into two groups termed "small copepods" and "large copepods." The species and stages within each group displayed similar vertical distributions. It was found, for example, that animals represented by the "large copepod" group migrate downward during day time whereas "small copepods" generally remain in the upper mixed layer.

Copepod distributions were analyzed for each geographical area (Table 9.1). Details of these analyses can be found in Herman (1984) for the Peru Shelf, Herman (1983) and Herman et al. (1984) for the Eastern Canadian Arctic, and Herman and Mitchell (1981) for the Scotian Shelf. Copepod groups (Table 9.1) for the Eastern Tropical Pacific were identified in a similar fashion.

OBSERVED VARIABILITY

Figure 9.2 illustrates 5 randomly selected chlorophyll profiles from Batfish transects in Peruvian coastal waters. The striking features of these profiles are the large differences in vertical structure, for example, in Figure 9.2(a, b) we see a surface and subsurface chlorophyll layer, a subsurface layer only in Figure 9.2(c), and mainly surface layers of varying concentration in Figure 9.2(d, e). Figure 9.3(a-d) illustrates 4 Batfish profiles from the Scotian Shelf consisting of chlorophyll, production, and copepods. Production profiles were generated from a chlorophyll/ light model (Herman and Platt, 1983; Herman and Platt, 1984) which has been verified with in situ incubation samples. Again we see large differences in vertical structure of all 4 parameters: chlorophyll, production, "small copepods," and "large copepods."

METHOD OF ANALYSIS

Horizontal Variability

We choose to estimate variability by considering a conservative property of each biological parameter,

179

TABLE 9.1.

Species and stages of copepods belonging to the "small copepods" and "large copepods" groups and separated into geographical areas. Staging where available is listed for most species.

	Peru	Eastern Tropical Pacific	Eastern Canadian Arctic	Scotian Shelf
Small copepods	Centropages brachiatus Paracalanus parvus Oncaea conifera Corycaeus giesbrichti Clausocalanus arculcornis	Oncaea venuta V, VI (F) Clausocalanus arculcornis Scolicithricella dentata VI (F) Euchaeta acuta IV	Pseudocalanus minutus Calanus hyperboreus III	Pseudocalanus minutus Clausocalanus arculcornis Meridia lucens V
Large copepods	Calanus chiliensis V, VI (M&F)	Eucalanus subtenuis IV, VI (F) Scolicitrix danae IV, VI (M&F) Euchaeta acuta V, VI (F) Undinnula daruimii V, VI (F) Sagitta sp. Thysanoessa sp. Euchaeta marina	Calanus finmarchicus V, VI (F) Calanus glacialis V	Calanus finmarchicus IV, V

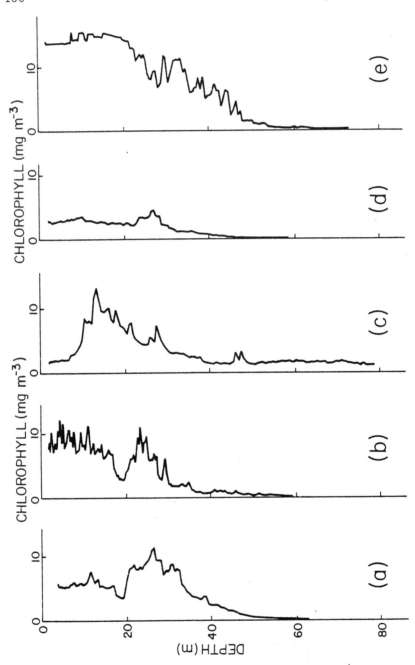

Figure 9.2. Randomly selected chlorophyll profiles sampled by Batfish in Peruvian coastal waters (a-e).

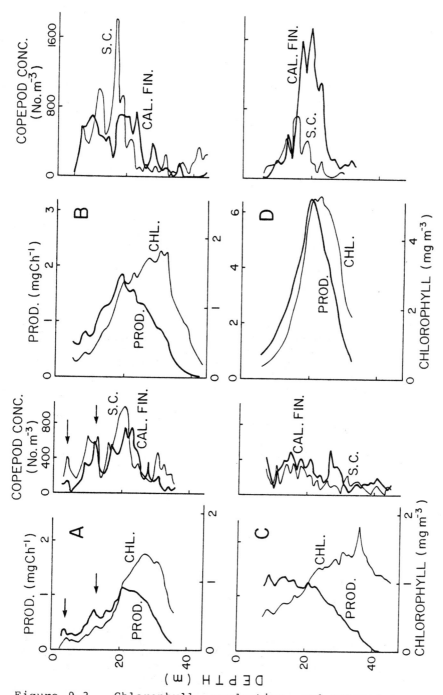

Figure 9.3. Chlorophyll, production, and copepod profiles sampled by the Batfish in Scotian Shelf waters (a-d). S.C. refers to small copepods.

that is, the integral of the parameter over depth. By
integrating over the water column within the euphotic
zone, we expect to avoid effects of vertical mixing,
internal waves, etc., while conserving total biomass.
If each parameter of chlorophyll, production, and copepods is represented by $Q(Z)$ as a function of depth Z,
we may then calculate the integral from:

$$QI = \int_{o}^{Z_f} Q(Z)dZ \tag{1}$$

when Z_f is a final depth chosen at 100 m, sufficient to
include all biomass in the euphotic zone. If we now
consider Batfish profile pairs separated by ΔX_1(km) as
shown in Figure 9.4, then the sample variance for a set
of successive profile pairs will be:

$$\sigma^2 (\Delta X_1) = \sum_{profile\ pairs} (QI_1 - QI_2)^2/N_p \tag{2}$$

where we sum over N_p, the numbers of profile pairs. We
may also estimate the variance at the next sample pair
separating ΔX_2, by summing over the number of successive pairs at a ΔX_2 separation as shown in Figure 9.4.
For the ith profile pair separation, we may then ex-
press the sample variance as:

$$\sigma_i^2 (\Delta X_i) = \sum_i (\delta QI)^2/N_p \tag{3}$$

The fractional standard error as a function of the
profile pair separation ΔX_1 may be written as:

$$FR(\sigma) = \sigma_i (\Delta X_i)/\overline{Q}(Z) \tag{4}$$

where $\overline{Q}(Z)$ represents the mean over the entire transect.

RESULTS: HORIZONTAL VARIABILITY

Peru

Figure 9.5(a) shows integrated chlorophyll and
integrated production plotted as a function of sampling
distance. The corresponding standard errors (eq. 4)
are also plotted in Figure 9.5(b) as a function of the
profile-pair sampling interval, ΔX_i.
Integrated production was high ranging from 2 to 3
g C m^{-2} d^{-1} and typical for the coastal waters of Peru

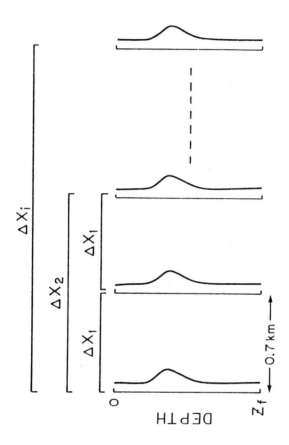

Figure 9.4. Successive Batfish profiles separated by a distance of 0.7 km denoted by ΔX_1. This represents the minimum sampling separation which will be compared statistically to larger separations such as ΔX_2 -----ΔX_i.

184

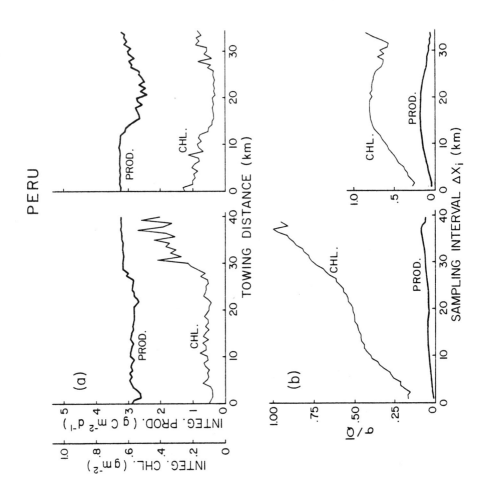

Figure 9.5.   Integrated chlorophyll and production (a)
plotted as a function of towing distance.   Lower plot
(b) illustrates the fractional sampling error relative
to either mean chlorophyll or mean production as a
function of sampling interval ΔX_i.

(Beers et al., 1971; Dagg et al., 1980; Walsh et al., 1980; Harrison and Platt, 1981). The standard error over sampling scales to 40 km was quite low, 9.6% (Table 9.2) averaged over 9 transects, and was seen to vary little (Figure 9.5[b]). The Peruvian production measurements were the lowest in variability among all the studied areas. This was mainly due to high concentrations of near-surface chlorophyll in Peruvian coastal waters and the subsequent restriction of production to the near-surface layer of 10-15 m depth (Herman, 1984). Integrated chlorophyll, on the other hand, showed considerably higher variability. Concentrations > 0.20 g m^{-2}, for example, measured at the end of the 1st transect (Figure 9.5[a]) were caused by a red tide event. The error curves of Figure 9.5(b) for both transects were very similar in shape for scales to 20 km while ranging in magnitude from about 20 to 100%. The large increase at scales > 20 km seen in Figure 9.5(b) for the 1st transect resulted from the red tide event. The mean standard error in chlorophyll averaged over 9 transects was 55% or nearly 5 times higher than that of production. It was shown by Herman (1984) that while production was restricted to the upper layers (0-15 m depth), most of the solar energy absorbed in the narrow depth layer was utilized for photosynthesis while resulting in a more constant integrated production for all sampling scales. There was a large fraction (2/3) of the standing stock measured in terms of chlorophyll a concentration, however, below this depth which was not active in photosynthetic production and it was this excess which provided the high variability in standing stock.

The error curve of Figure 9.5(b) suggests that there exists a characteristic length scale at which point small scale turbulent diffusion is balanced by phytoplankton growth. The error curves for the second transect increase in a near linear fashion up to 10 or 12 km at which point it levels out. In the first transect, the error curve again increases about 20 km due to the red tide event. This "knee" at 10 km appears only weakly in the production error curves. This was expected since production was restricted to the upper 15 m; whereas most of the chlorophyll biomass was measured at the thermocline where the error curve was most sensitive.

Integrated copepods (# m^{-2}) for the same Batfish transects and their corresponding error curves are shown in Figure 9.6(a,b). The mean standard error for 9 transects (Table 9.2) indicated higher variability (≈ 87%) than either chlorophyll or production as, we shall see, is the case for nearly all geographical areas. There were no visible trends among the data nor any differences in day and night tows. There was an

186

TABLE 9.2.

The mean standard error expressed as a percentage for the four parameters as a function of each geographical sampling area. Numbers in brackets beside each location represent the number of transects used in averaging the mean error.

Parameter Q (z)	Peru (9)	East. Trop. Pac. (3)	Scotian Shelf (3)	Baffin Bay (4)	Lancaster Sound (9)
			Mean Standard Error (%)		
Chlorophyll	55.0	8.0	42	42	71
Production	9.6	7.8	20	45	86
Small copepods	88.0	45.0	52	55	97
Large copepods	86.0	33.0	71	45	59

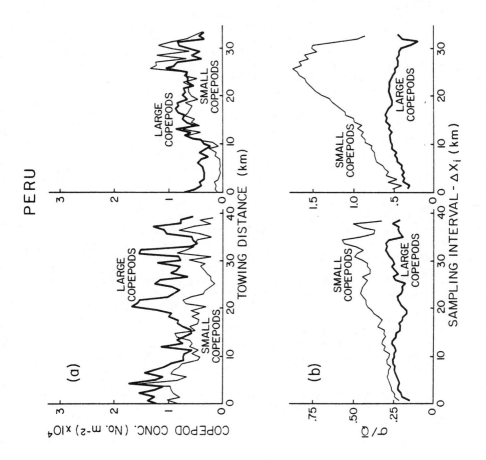

Figure 9.6. Integrated copepod concentrations (a) as a function of towing distance. Sampling errors for small and large copepods are plotted (b) as a function of sampling error ΔX_i.

increase in error as a function of sampling interval.
Linear least-squares fits yielded a mean intercept
of ≈ 40% indicating that copepod variability is badly
undersampled. The mean slope indicated an increase of
≈ 3% per km. There was a "knee" at 10 km in the chlorophyll error curve while none was observed in the
copepod error curves. This difference may be attributed to migrational patterns observed (Herman, 1984).
Small copepods resided in the upper mixed layer from 0-
20 m both day and night; however, some migration was
observed within this strata. Large copepods clearly
resided in the mixed layer during nighttime and migra-
ted to a maximum depth of 100 m during daytime.

Eastern Tropical Pacific

    Figure 9.7(a,b) illustrates the integrated chlo-
rophyll, production, and their corresponding error
curves. Sampling methods on this cruise did not provide sufficient data suited for these analyses; transects were repeated along the same line and were only
< 20 km in length. As one might expect from deep ocean
data, variability was quite low while the standard
error for both chlorophyll and production was only
about 8% over all scales. Sampling errors for the
Peruvian production data were low as a result of high
surface concentrations of chlorophyll; however, this
was not the case for the Tropical Pacific since both
production and chlorophyll maxima were deep (≈ 30 to 50
m). The low sampling errors were due to the vertical
constancy of the profiles. There was no discernible
"knee" observed in Figure 9.7(b).
    Integrated copepods and their corresponding error
curves are shown in Figure 9.7(c,d). The mean errors
(Table 9.2) were the lowest among all the marine ecosystems ranging from 30 to 45%. Although Figure 9.7(d)
exhibited an increase in error with sampling interval,
this was not a consistent feature in other tows which
demonstrated a zero slope. No apparent "knee" was
observed at scales < 15 km.

Scotian Shelf

    Figure 9.8(a,b) represents chlorophyll, produc-
tion, and their corresponding error curves for two
Batfish transects. The second transect represents a
reverse course of the first made approximately 4 hours
apart. The mean standard error for chlorophyll was 42%
(Table 9.2) while production variability was about a
factor of 2 lower, although still greater than production errors for the Peruvian and Eastern Tropical
Pacific data.

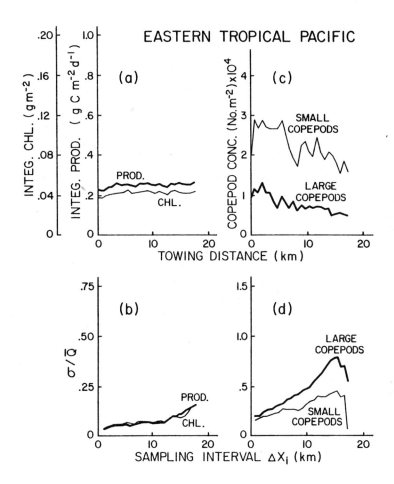

Figure 9.7. Integrated chlorophyll and production (a) and their corresponding error curves (b) plotted as a function of towing distance and sampling interval, respectively. Integrated copepods (c) and error curves (d) plotted as a function of towing distance and sampling interval, respectively.

190

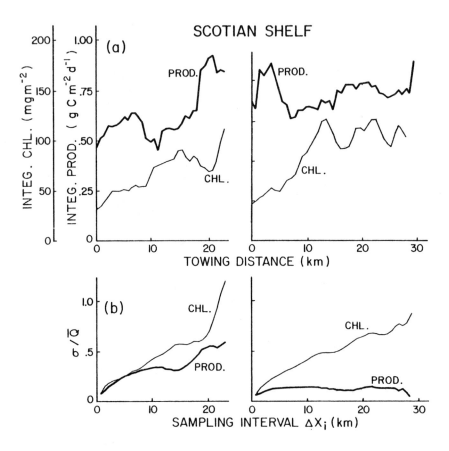

Figure 9.8. Integrated chlorophyll and production (a) and their corresponding error curves (b) for two Batfish transects. Right hand plot represents reverse course of the same transect.

The transects were sampled at the outer edge of the Scotian Shelf in a frontal region dominated by M_2 internal tides (Herman and Denman, 1979). In the first transect, while moving from left to right, chlorophyll and production, in particular, increased while nearing the shelf break front. The subsurface chlorophyll maxima was at about 30 m in depth while the production maxima occurred at about 20-m depth. Nearing the frontal outcropping at the end of the transect, these depths were approximately halved. The production increase, however, was independent of the increase in integrated chlorophyll and due to a shallower subsur-face layer of chlorophyll near the front enabling plant cells to utilize solar energy more efficiently (Herman et al., 1981). Closer examination of integrated chlorophyll and production revealed that small scale changes (2 to 4 km) were actually out of phase.

In the second transect, we see changes in the integrated chlorophyll and production caused by the M_2 internal tides. The integrated chlorophyll at the front is reduced by a mixing event from below (Herman and Denman, 1979), yet there was sufficient near-surface chlorophyll to maintain the same production. Note that the overall sampling error in production is more than halved.

Figure 9.9(a,b) represents the same Batfish transects consisting of large and small copepods and corresponding error curves. Both transects showed large error peaks at 10 km corresponding to "small copepods" peaks in Figure 9.9(a). The mean sampling error for large copepods approximately doubled in the second transect while indicating peak sampling errors on nearly all sampling scales up to 30 km. Only one "knee" at 3 km was observed for large copepods in the second transect.

Baffin Bay

Chlorophyll, production, copepods, and their corresponding error curves are shown in Figures 9.10(a,b) and 9.11(a,b). The significant feature that we observe in all these data is that the error curves are reasonably constant over all scales and the mean standard error was about 45% for all parameters--chlorophyll, production, and both copepod groups. This can be understood by examining the vertical structure of the Baffin Bay profiles described by Herman (1983). The thermocline was at about 70 m depth, narrow, and had little horizontal variability. Chlorophyll was situated at the thermocline and also narrowly layer-ed. Both production and copepods were also narrowly layered while their mean depths were about 3 to 5 m above the chlorophyll layer.

192

SCOTIAN SHELF

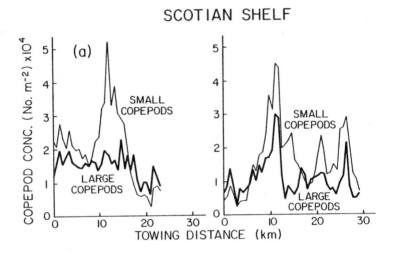

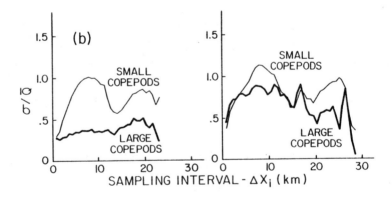

Figure 9.9. Integrated copepods (a) and their corresponding error curves (b) for two Batfish transects. Right hand plot represents reverse course of the same transect.

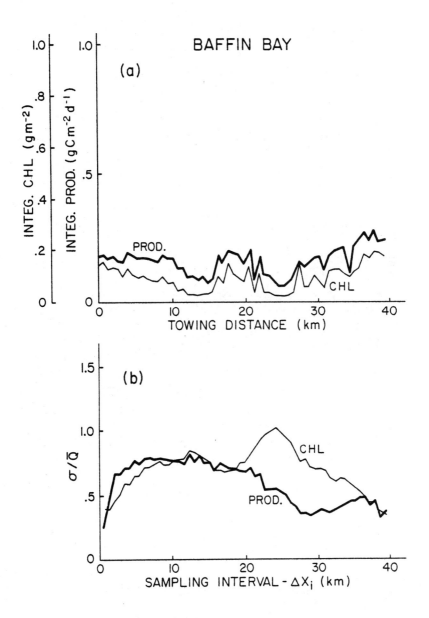

Figure 9.10. Integrated chlorophyll and production (a), and their corresponding error curves (b).

194

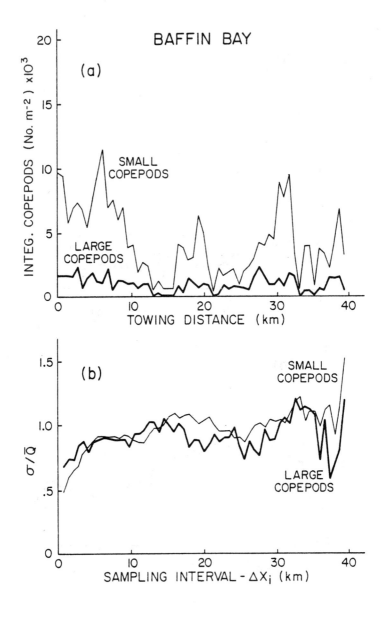

Figure 9.11.   Integrated copepods (a) and their corresponding error curves (b).

Both copepod groups did not undergo diel migration
although small vertical changes ($\approx$ 5 m) were observed
within the mixed layer. All four parameters exhibited
similar vertical structure and therefore similar sampling error. A "knee" at about 4 to 5 km was observed
on the error curves of all parameters.

Lancaster Sound

The four parameters and their error curves are
shown in Figures 9.12(a,b) and 9.13(a,b). Lancaster
Sound exhibited the largest sampling error of all the
marine ecosystems sampled. This was caused, for the
most part, by its physical oceanographic features. The
current structure (Milne and Smiley, 1978) at the mouth
of the Sound consists of Baffin Bay water entering the
north side, flowing west about 50-100 n.m., and
subsequently swinging south across the Sound. The current then changes to an easterly direction along the
south side of the Sound and, before exiting into Baffin
Bay, admixes with Arctic Ocean currents. The sampled
mixed layer depth (Herman et al., 1984) was highly
variable while ranging from 80 to 100 m for some
portions of the transect to 20 m for others. As a
result, chlorophyll found near the surface resulted in
higher production as seen on either side of the Sound
(Figure 9.12(a)) whereas the production in the middle
of the Sound was decreased by 8 times.

The error curve for production (Figure 9.12(b))
presents an interesting problem in sampling. The same
low errors of about 25% can be achieved by sampling
either at small scales of 1 to 2 km along the entire
transect or at larger scales of 45-50 km using only a
few stations. If such horizontal structure was a
persistent feature, then the mean production could be
easily determined from only a few stations.

Copepods (Figure 9.13(a)) near the mouth of the
Sound were evenly distributed throughout the upper 100
m depth and there was no layering observed near the
surface. The error curves (Figure 9.13(b)) were nearly
constant over all sampling scales. The only "knee"
observed among all four parameters was in the
chlorophyll error curve at about 4 km.

DISCUSSION

Overall compositional structure, with the
exception of the two copepod size groups was not
measured in this study. In the Scotian Shelf data,
small copepods (Figure 9.9) consisting mostly of
Pseudocalanus minutus and Clausocalanus arcuicornis
exhibited two (possibly more) distinct error peaks or
"nodes" corresponding to aggregations along a frontal

196

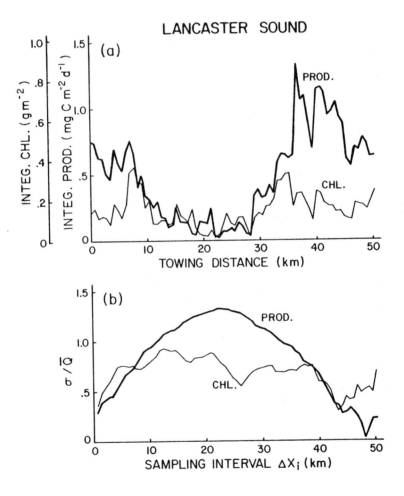

Figure 9.12. Integrated chlorophyll and production (a), and their corresponding error curves (b).

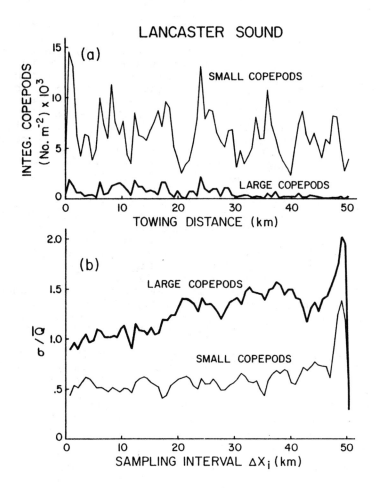

Figure 9.13.   Integrated copepods (a) and their
corresponding error curves (b).

system located at the Scotian Shelf break. The error measured for large copepods was nearly constant over all scales up to 30 km. Physical changes at long times scales of ≈ 1 month (Herman and Denman, 1979) suggests that such large spatial scale aggregations of small copepods are associated with behavioral aggregations such as migration or a response to the temperature front which acts as a barrier. Both error curves for chlorophyll and production (Figure 9.8) increased linearly with the sampling interval as a direct result of advective control. In an onshore to offshore direction, chlorophyll concentrations increased near the shelf break front as a result of intense vertical mixing of nutrients from underlying slope water (Herman and Denman, 1979). Therefore, on large spatial scales, one finds large sampling errors for the integrated standing stock. Production on the other hand is most sensitive to the vertical structure of chlorophyll which may vary considerably in the shelf break front. Chlorophyll layers are often "elevated" near the surface at the frontal system which results in significant production increases. These "elevations" caused by the front occur on short time scales (i.e., semidiurnal tidal cycles ≈ 1/2 day) resulting in large variations in the error curves over large spatial scales (≈ 30 km) even for the two transects in Figure 9.9 which were sampled in a period of less than 12 hours.

In the case of Lancaster Sound, advective changes caused by intrusions of Baffin Bay water occurred on short time scales of ≈ 1 day resulting in an alternately broad or sharp thermocline gradient over the 50-km transect. Error curves for both large and small copepods (Figure 9.13) and chlorophyll (Figure 9.12) were nearly constant over all sampling scales up to 50 km, suggesting strong advective control in dispersing behavioural aggregations. Production on the other hand, exhibited large variability (Figure 9.12) which was due primarily to variations in the vertical scale in the vertical chlorophyll layers on scales of ≈ 20 km resulting in enhanced production.

In a sharply stratified system where plants, animals, and production coexist within a narrow depth strata, one might expect to find similar magnitudes of sampling errors which remain constant over all sampling scales. This result was most pronounced in the Baffin Bay data (Figures 9.10-9.11), particularly in the case of large and small copepods. Production errors (Figure 9.10(b)) were not constant at large spatial scales of ≈ 30-40 km due to the variable presence of glacial material (non-photosynthetic) in the upper 10 m, thereby attenuating light and reducing water column production. Notably there appeared a "knee" at about 2-3 km

in all error curves, indicating finer scale physical
control of growth and aggregations.

Previous investigations of geographic and temporal
variability of zooplankton and phytoplankton (concentration per m^{-2}) were made (Mackas and Sefton, 1982;
Mackas, 1984) in British Columbian coastal waters.
They observed a strong spatial autocorrelation of the
plankton community structure over length scales of
order 50 km. The temporal variability of their
community pattern for both zooplankton and phytoplankton was low. Compositional patches were coherent over
time scales of 1-20 days and over space scales of 50
km, but exhibited large compositional changes over
larger time intervals. Changes in their local compositional structure were found to be under advective
control. Similar evidence has been found for the
Oregon coast where short time scales (1-15 days) variations in zooplankton composition were attributed
(Peterson et al., 1979; Smith et al., 1981) to variations in the onshore-offshore component of flow.
Photosynthetic measurements of Mackas and Sefton (1982)
showed no dependence of assimilation number (mg C (mg
chl a)$^{-1}$ h^{-1}) on chlorophyll or nitrate concentration. The Batfish sampling encompassed scales from
≈ 0.5-30 km; whereas, their sampling scales ranged
from ≈ 10-200 km; therefore, our data are only comparable for larger spatial scales, or 10-30 km. We are
unable to test the zooplankton community structure at
spatial scales of 50 km; however, it was apparent that
much of our error data were also dependent on advective
control at lower spatial scales. We also concur with
their findings of no dependence of assimilation number
on chlorophyll concentration while we also found no
distinct dependence between integrated production and
integrated chlorophyll.

REFERENCES

Barnes, H., and Marshall, S. M. 1951. On the varia-
    bility of replicate plankton samples and some
    applications of contiguous series to statistical
    distribution of catches over restricted periods.
    J. Mar. Biol. Ass. U.K. 30:233-262.
Beers, J. R., Stevenson, M. R., Eppley, R. W., and
    Brooks, E. R. 1971. Plankton populations and
    upwelling off the coast of Peru, June 1969. Fish.
    Bull., U.S. 69:859-876.
Dagg, M., Cowles, T., Whitledge, T., Smith, S., Howe,
    S., and Judkins, D. 1980. Grazing and excretion
    by zooplankton in the Peru upwelling system during
    April 1977. Deep-Sea Res. 27:43-59.

Denman, K. L., and Platt, T. 1975. Coherences in the horizontal distributions of phytoplankton and temperature in the upper ocean. In Proceedings of the sixth Liege colloqium on ocean hydrodynamics. Ed. by J. Nihoul. Mem. Soc. R. Sci. Liege, 6^e series 7:19-30.

Dessureault, J.-G. 1976. "Batfish": A depth controllable towed body for collecting oceanographic data. Ocean Eng. 3:99-111.

Fasham, M. J., and Pugh, P. R. 1976. Observations on the horizontal coherence of chlorophyll a and temperature. Deep-Sea Res. 23:527-538.

Hardy, A. C. 1956. The open sea: Its natural history part 1. The world of plankton. Collins, St. James' Place, London. 335 pp.

Harrison, W. G., and Platt, T. 1981. Primary production and nutrient fluxes off the northern coast of Peru. A summary. Bol. Inst. Mar Peru (Callao). pp. 15-21.

Hasle, G. R. 1954. The reliability of single observations in phytoplankton surveys. Nytt. Mag. Bot. (Oslo) 2:121-137.

Herman, A. W. 1982. Spatial and temporal variability of chlorophyll distributions and geostrophic estimates on the Peru shelf at 9° S. J. Mar. Res. 40:185-207.

Herman, A. W. 1983. Vertical distribution patterns of copepods, chlorophyl, and production in northeastern Baffin Bay. Limnol. Oceanogr. 28:709-719.

Herman, A. W. 1984. Vertical copepod aggregations and interactions with chlorophyll and production on the Peru shelf. Continental Shelf Res. (In press).

Herman, A. W., and Dauphinee, T. M. 1980. Continuous and rapid profiling of zooplankton with an electronic counter mounted on a "Batfish" vehicle. Deep-Sea Res. 27:79-96.

Herman, A. W., and Denman, K. L. 1976. Rapid underway profiling of chlorophyll with an in situ fluorometer mounted on a Batfish vehicle. Deep-Sea Res. 24:385-397.

Herman, A. W., and Denman, K. L. 1979. Intrusions and vertical mixing at the shelf-slope water front south of Nova Scotia. J. Fish. Res. Board Can. 36:1445-1453.

Herman, A. W., and Mitchell, M. R. 1981. Counting and identifying copepod species with an in situ electronic zooplankton counter. Deep-Sea Res. 28:739-755.

Herman, A. W., and Platt, T. 1983. Numerical modelling of diel carbon production and zooplankton grazing on the Scotian Shelf based on observational data. Ecol. Model. 18:55-72.

Herman, A. W., and Platt, T. 1984. Primary production
profiles in the ocean: Estimation from a
chlorophyll/light model. (In manuscript).

Herman, A. W., Sameoto, D. D., and Longhurst, A. R.
1981. Vertical and horizontal distribution in
patterns of copepods near the shelf break south of
Nova Scotia. Can. J. Fish. Aquat. Sci. 38:1065-
1076.

Herman, A. W., Sameoto, D. D., and Longhurst, A. R.
1984. Vertical distributions of copepods,
chlorophyll, and production in Lancaster Sound.
(In manuscript).

Kierstead, H., and Slobodkin, L. B. 1953. The size of
water masses containing plankton blooms. J. Mar.
Res. 12:141.

Lekan, J. F., and Wilson, R. E. 1978. Spatial variability of phytoplankton biomass in the surface
waters of Long Island. Estuarine Coastal Mar.
Sci. 6:239-251.

Mackas, D. L. 1984. Spatial autocorrelation of
plankton community composition in a continental
shelf ecosystem. Limnol. Oceanogr. 29(3):451-471.

Mackas, D. L., and Sefton, H. A. 1982. Plankton
species assemblages off southern Vancouver Island:
Geographic pattern and temporal variability. J.
Mar. Res. 40(4):1173-1200.

Milne, A. R., and Smiley, B. D. 1978. Offshore dril-
ling in Lancaster Sound: Possible environmental
hazards. Dep. Fish. and Environ. Report, Inst.
Ocean Sciences, B.C.

Peterson, W. T., Miller, C. B., and Hutchinson, A.
1979. Zonation and maintenance of copepod populations in the Oregon upwelling zone. Deep-Sea
Res. 26:467-494.

Platt, T. 1972. Local phytoplankton abundance and
turbulence. Deep-Sea Res. 19:183-187.

Powell, T. M., Richerson, P. J., Dillon, T. M., Agee,
B. A., Dozier, B. J., Godden, D. A., and Myrup,
L. O. 1975. Spatial scales of current speed and
phytoplankton biomass fluctuations in Lake Tahoe.
Science 189:1088-1090.

Smith, S. L., Brink, K. H., Sanlander, H., Cowles,
T. J., and Huyer, A. 1981. The effect of advec-
tion on variation in zooplankton at a single loca-
tion near Cubo Nazca, Peru, in Coastal Upwelling.
Ed. by F. A. Richards. Am. Geophys. Union. Wash.,
D.C. 529 pp.

Steele, J. H. 1976. Patchiness. In The ecology of
the seas. pp. 90-115. Ed. By D. H. Cushing and J.
J. Walsh. Blackwell Scientific Publications,
Oxford. 467 pp.

Steele, J. H., and Henderson, E. W. 1977. Plankton
patches in the northern North Sea. In Fisheries

202

mathematics.   pp. 1-19.   Ed. by J. H. Steele.
Academic Press, London.
Walsh, J. J., Whitledge, T. E., Esaisas, W. E., Smith,
R. L., Huntsman, S. A., Santander, H., and
DeMendiola, B. R. 1980. The spawning habitat of
the Peruvian anchovy, *Engraulis ringens*. Deep-Sea
Res. 27:1-27.
Wiebe, P. H., Burt, K. H., Boyd, S. H., and Morton, A.
W. 1976. A multiple opening/closing net and en-
vironmental sensing system for sampling zooplank-
ton. J. Mar. Res. 34:313-376.

10. Measurement Strategies for Monitoring and Forecasting Variability in Large Marine Ecosystems

ABSTRACT

The large marine ecosystems (LMEs) within the newly designated Exclusive Economic Zone of the United States are characterized by unique bathymetry, hydrography, productivity, and trophically linked population structure. Significant effort is underway aimed at providing a scientific basis for the management and conservation of living resources within seven LMEs--the Insular Pacific, Eastern Bering Sea, Gulf of Alaska, California Current, Gulf of Mexico, Southeast Atlantic Shelf, and Northeast Atlantic Shelf. In each of the LMEs three resource assessment strategies have been implemented to monitor variability and forecast abundance of resource populations: (1) utilization of yield statistics for estimating population trends, (2) yield-independent surveys of adult and early-life stages on mesoscale spatial (20-100 km) and temporal (weeks-months) sampling frequencies, and (3) process-oriented studies of ecosystem structure and function leading to improved resource forecasts. For LMEs in which a time series of these kinds of measurements have been made, trade-off options for achieving optimum resource yields are being developed.

INTRODUCTION

Resource assessment studies of the National Marine Fisheries Service (NMFS) were expanded significantly during the middle 1970s to support the conservation and management of marine fishery resources within the U.S. Fisheries Conservation Zone (FCZ) established by Congress in 1976. This law extended U.S. jurisdiction to a 322 km (200-mile) wide strip of ocean off all the U.S. coasts (over 3.5 million km^2). The region was redesignated as the Exclusive Economic Zone (EEZ) in 1983 by presidential proclamation. In this report an

203

overview is provided of the research strategies and new studies implemented by NMFS to overcome the problems posed by the large-scale temporal and spatial biological and environmental changes influencing the abundance levels of U.S. fishery resources within the EEZ.

The studies are part of a NMFS/NOAA multispecies and environmental initiative known as the Marine Resources Monitoring, Assessment, and Prediction (MARMAP) program. The MARMAP program was built around a matrix of existing NMFS fishery resource assessment activities including studies dealing with the analyses of yield (catch) statistics, the results of fishery surveys (pelagic, demersal, ichthyoplankton), fisheries oceanography, and fisheries engineering. A full description of the MARMAP assessment system elements is given in a series of planning documents prepared by NMFS with the assistance of the Ocean Systems Division of TRW Company[2] (1973a,b, 1974). The coordination and integration of investigational components of the assessment system is a major research activity of the four Fisheries Centers of NMFS. The Northwest and Alaska Fisheries Center, Seattle, Washington, is responsible for studying resources in the Gulf of Alaska, Eastern Bering Sea, and off the coasts of Washington and Oregon. The Southwest Fisheries Center, La Jolla, California, has responsibility for the studies of the living resources of the California Current, Hawaii, and the Pacific Trust territories. The Southeast Fisheries Center, Miami, Florida, assesses the resources from North Carolina to the Florida Keys, and in the Gulf of Mexico and Caribbean. The Northeast Fisheries Center, Woods Hole, Massachusetts, studies the resources on the continental shelf from the Gulf of Maine to Cape Hatteras. The unique, energetically-related biological communities, bathymetry, hydrography, circulation, and productivity within each of these regions comprise coherent ecological systems encompassing broad geographic areas designated as LMEs (Figure 10.1).

The fishery resources within the LMEs are subject to management by Regional Fisheries Management Councils. Their management plans are directed toward ensuring optimal sustained yields based on ecological, economic, and social considerations. The ecological decisions are based on the best scientific information available. Each Center conducts multispecies and environmental assessment studies to support the Councils in developing management and conservation plans for regional fishery resources. In addition to direct support of fisheries management, NMFS also provides information on the status of the environment (habitat) to the management councils and other federal agencies (e.g., EPA, Corps of Engineers, Dept. of

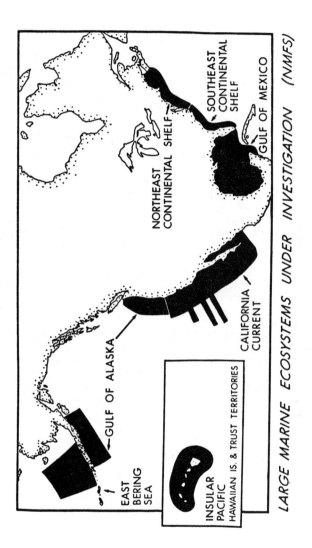

Figure 10.1. Large marine ecosystems (LMEs) where
fishery stock assessment studies are underway by the
National Marine Fisheries Service. (Adapted from
Sherman et al., 1983b).

Interior) mandated to regulate the utilization of resources and ocean habitat.

FISHERIES STUDIES IN LMEs

From the turn of the century through the middle
1970s, fisheries studies were mainly focused on the
yields of single species. From a fisheries management
point of view, the best and most sought after scientific information is an accurate prediction of future
stock sizes and the effects of different levels of
fishing or environmental perturbation on the continued
production of economically viable resource populations.

At present, NMFS is developing a holistic approach
to fishery assessment studies with a focus on whole
ecosystems and the multispecies interactions at different trophic levels that influence the annual production of fish populations. A balanced approach is
being implemented by NMFS to obtain the comprehensive
population and environmental information required to
improve forecasts of fish abundance within the EEZ that
allows for:

(1) a time series of measurements in the form of
standardized multispecies resource assessment and
hydrographic surveys,

(2) a systematic collection of fish-catch data
(yields), and

(3) process-oriented studies dealing with biological and environmental linkages and energetic
transfers among key ecosystem components important to
fish production in the sea.

Studies of single species alone do not provide
sufficient data for effective management of multispecies fisheries operating at different trophic
levels. While it is important for management purposes
to continue these studies, they are now being pursued
by NMFS within a broader matrix that measures interactions that may result in changing abundance levels
among the key species in the ecosystem. Single-species
yield simulation models have been augmented with more
holistic multispecies models (Regier and Henderson,
1973; Parrish, 1975; Sheldon et al., 1977; Andersen and
Ursin, 1977; Jones, 1976; Sherman et al., 1978;
Laevastu and Favorite, 1978, 1981; Beddington et al.,
1979; Grosslein et al., 1980; Laevastu and Larkins,
1981; Mann, 1982; Sissenwine et al., 1984). These
models deal with multispecies fishery interactions at
different trophic levels. They are important approxi-
mations of the consequences of predator-prey dynamics,
based on fishery imposed selective mortality, and hold
promise for providing a basis for the management of
marine ecosystems.

In the LMEs, several assessment strategies are used by NMFS to improve abundance forecasts of recruitment success of incoming year classes. Fisheries-independent surveys of fish eggs and larvae are conducted on mesoscale grids of 20-100 km at frequencies of two to twelve times a year to estimate the size of the spawning adult stocks. Ichthyoplankton surveys represent an effective sampling strategy for measuring abundance levels of multispecies fish communities inhabiting LMEs. The CalCOFI (California Cooperative Oceanic Fisheries Investigations) studies pioneered by Ahlstrom (1954) attest to the tractability of measuring population level changes in the ichthyoplankton of the California Current system. The CalCOFI prototype ichthyoplankton survey was used as the standard approach in the multispecies assessments and adapted for use in the LMEs under investigation by NMFS. The eggs and larvae of most marine fish species in an LME can be quantitatively sampled with a single device--the plankton net. The early developmental stages are all vulnerable to the paired 60-cm bongo nets used on NMFS surveys (Posgay and Marak, 1981). Trawl surveys and pelagic surveys employing net systems are also used to assess population levels of prerecruit and adult stages of fish stocks. However, they are more selective samplers. The sampling designs of the multispecies ichthyoplankton surveys within the LMEs provide measures of spatial and temporal variability that are within acceptable confidence limits for estimating trend changes in abundance levels of parental spawning biomass (Stauffer and Charter, 1982; Pennington and Berrien, 1982). To obtain samples of ichthyoplankton used in spawning biomass estimates, the sampling is designed to encompass the temporal and spatial extent of spawning using a systematic grid of stations. A detailed description of the methods used by NMFS for multispecies ichthyoplankton sampling is given in Smith and Richardson (1977) and Jossi and Marak (1983).

Within the mesoscale (20-100 km) multispecies ichthyoplankton time-series surveys (bimonthly to semiannual), studies of the recruitment process are conducted for target species on a finer horizontal and vertical scale (Lasker, 1981a; Lough and Laurence, 1982) aimed at discovering the processes controlling annual recruitment success of new year classes. Processes under investigation include growth and mortality of eggs, larvae, and juveniles under variable density dependent predator-prey interactions and density independent influences of changes in circulation, water-column structure, biological production, and pollution. Among the target species of recruitment studies are walleye pollock, *Theragra*

208

chalcogramma; Pacific king crab, Paralithodes; tanner crabs, Chionoectes; Pacific sardine, Sardinops; Pacific anchovy, Engraulis mordax; Atlantic mackerel, Scomber scombrus; Pacific salmon, Oncorhynchus; striped bass, Morone saxatilis; Pacific hake Merluccius productus, silver hake Merluccius bilinearis, Atlantic menhaden, Brevoortia tyrannus, Gulf shrimp, Penaeus; bluefin tuna, Thunnus thynnus; spot, Leiostomus xanthurus; Atlantic croaker, Micropogonias undulatus; Atlantic cod, Gadus morhua; and haddock, Melanogrammus aeglefinus.

PROTOTYPE STUDIES: CALIFORNIA CURRENT

The NMFS Southwest Fisheries Center (SWFC) has an extensive larval fish program in its Coastal Fisheries Resources Division. Studies on larval fish at the SWFC were begun by the late Elbert H. Ahlstrom who is acknowledged as a pioneer in larval fish identification and as the originator of egg and larvae surveys to determine the distribution and number of fish in the sea (Ahlstrom, 1954, 1959, 1965). The areal extent of the CalCOFI ichthyoplankton studies in the California Current ecosystem is shown in Figure 10.2. The ichthyoplankton abundance information obtained by Ahlstrom was instructive in documenting the failure of sardine recruitment in the California Current ecosystem and the increase in anchovy biomass, particularly in the absence of commercial fishery and associated catch statistics for the anchovy stocks during the 1950s and 1960s (Ahlstrom, 1966; Kramer and Smith, 1971) (Figure 10.3).

Recruitment studies conducted by Lasker and his group at the Southwest Fisheries Center, La Jolla, California, have demonstrated the importance of the intensity of upwelling to the growth and survival of anchovy larvae in relation to the abundance of their dinoflagellate prey (Lasker, 1981b). Year-class strength of anchovies may depend on the presence of Gymnodinium splendens in contrast to Gonyaulax polyedra and on the maintenance of moderate levels of layering within the California Current ecosystem (Lasker, 1981b).

STRESSED NORTHEAST SHELF ECOSYSTEM

The continental shelf ecosystem off the U.S. northeast coast supports a fisheries industry that contributes one billion dollars annually to the economies of the coastal states from Maine to North Carolina. Since 1963, the NEFC has conducted bottom-trawl surveys over the entire shelf ecosystem from the Gulf of Maine to Cape Hatteras. A systematic time

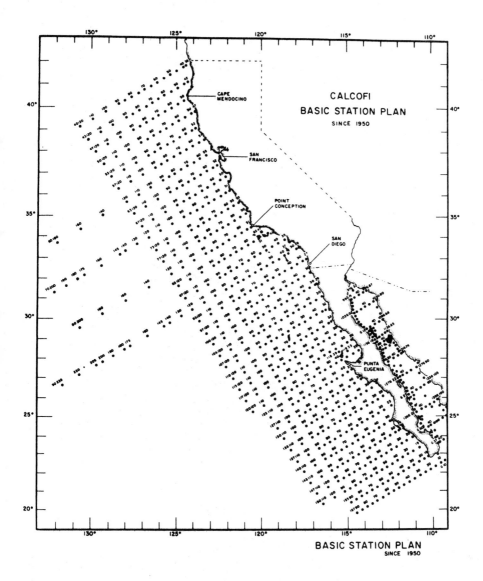

Figure 10.2. CalCOFI ichthyoplankton station pattern since 1950.

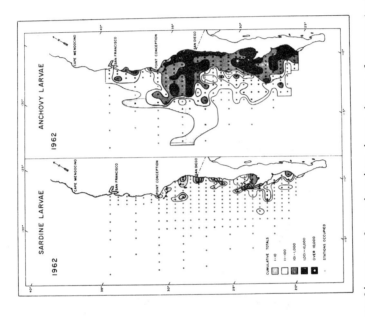

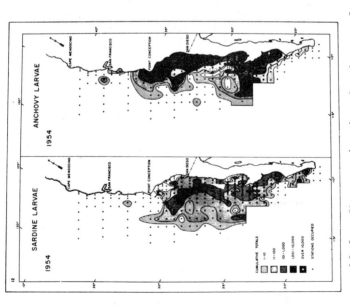

Figure 10.3. Changes in the abundance of sardines and anchovies have been documented through the CalCOFI ichthyoplankton survey method. (a) Relative abundance of sardine and anchovy larvae in 1954. (b) The decline in sardines and population increase in anchovy based on CalCOFI surveys in 1962. (From Ahlstrom, 1966).

series of yield statistics from commercial fishing operations have also been collected during the past 20 years. The fish stocks of the region have been heavily exploited. From 1968 through 1975 the total catchable finfish biomass declined by approximately 50% (Figure 10.4). This decline was correlated with high fishing mortality (Clark and Brown, 1977). Since 1975, a slight recovery trend has been observed among the demersal species (i.e. cod, Gadus morhua; pollock, Pollachius virens, flounders, Paralichthys, Hippoglossoides, Limanda, and Pseudopleuronectes). Atlantic herring, Clupea harengus and Atlantic mackerel stocks remain depressed (Brown et al., 1984).

The dramatic decline raised several important questions. Would the reduction of predation pressure by the loss of pelagic zooplanktivorous fish result in elevated levels of zooplankton? Would small, fast-growing, opportunistic zooplanktivorous species replace the herring and mackerel populations? Would the depressed stock return to former abundance levels with the control of fishing mortality imposed by the establishment of the FCZ/EEZ and the significant reduction of large-scale factory-trawler operations? In an effort to address these questions, surveys of ichthyoplankton were expanded in 1977 to cover the northeast shelf ecosystem and provide fisheries-independent information on the total ichthyoplankton community of the system. Following the CalCOFI model, a systematic network of sampling for ichthyoplankton was established on a grid network with stations spaced 25-30 km apart over the entire 260,000 km^2 of the northeast shelf (Figure 10.5). At each station collections were made with paired bongo nets fitted with 0.333-mm and 0.505-mm mesh nets. In addition, water column sampling was conducted for temperature, salinity, nutrients, oxygen, chlorophyll, and primary production (^{14}C) (Evans and O'Reilly, 1983; O'Reilly and Thomas, 1983). An average of 6 surveys were made each year from 1977 through 1981. All ichthyoplankton and zooplankton collections were sent to the Polish Sorting Center in Szczecin, Poland, for processing.

The 1977-1981 MARMAP surveys provided new information on the productivity of the shelf ecosystem. The shelf ecosystem was divided into four subareas based on areal differences in bathymetry, hydrography, circulation, and population structure--Gulf of Maine, Georges Bank, Southern New England, and the Mid-Atlantic Bight. With the exception of the shelf-slope front, the shelf ecosystem is highly productive. Mean annual values of carbon production ranged from 260 g C/m^2 in the mid-shelf off Cape Hatteras to 455 g C/m^2 on Georges Bank (O'Reilly and Busch, 1984). For the Georges Bank area interannual variability in chlorophyll biomass for a

212

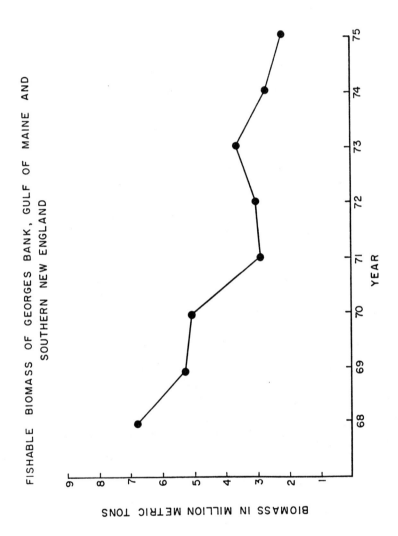

Figure 10.4. Decline in the fishable biomass of Georges Bank, Gulf of Maine, and Southern New England between 1968 and 1975. (Adapted from Clark and Brown, 1977.)

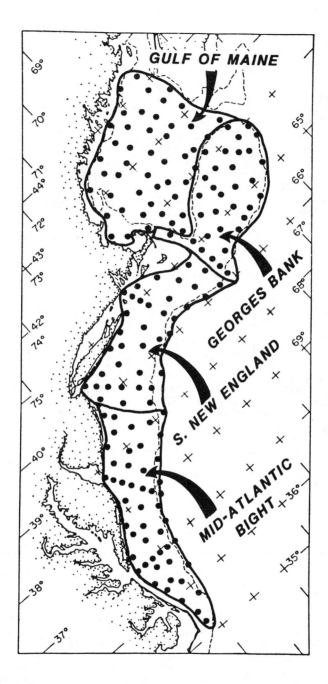

Figure 10.5.   Station locations for sampling ichthyoplankton on the northeast continental shelf ecosystem.

6-yr time series 1977 to 1982 ranged from 16% to 64% (Table 10.1). Also, a comparison between contemporary MARMAP ^{14}C values with primary productivity measurements made in the 1940s were similar (Table 10.2). It appears from these observations that the northeast shelf ecosystem is far less variable in primary production than other areas including the southern North Sea and in the upwelling region off Peru where mean annual productivity can differ by three orders of magnitude (Cushing, 1983). Mean annual primary productivity is also highly variable in the California Current system (Figure 10.6).

Recurrent annual cycles of zooplankton abundance within the northeast ecosystem were observed from 1977 through 1981 in each of the four subareas (Sherman et al., 1983a). Displacement volumes expressed as cc/100 m^3 of water strained are used to represent standing stocks of zooplankton. The seasonal patterns of zooplankton biomass observed each year and compared with the 5-yr means in each of the subareas, were coherent (Figure 10.7). The term coherent is used here to describe the recurring temporal patterns of zooplankton biomass in which annual deviations from the 5-yr mean are insignificant at the 0.05 level (Table 10.3). Over the 5-yr time series the spatial and temporal patterns of the dominant zooplankters were similar in each of the four subareas (Figure 10.7). The contemporary mean seasonal zooplankton dominance and standing stock abundances were not significantly different from earlier studies of zooplankton made during the first half of the century (Sherman et al., 1983a) (Table 10.4). Unlike the changes in zooplankton reported for the California Current ecosystem (Wickett, 1967; Colebrook, 1977) and the Northeast Atlantic (Colebrook, 1978a,b) the plankton of the northeast shelf ecosystem does not appear to have changed substantially over the past 70 years.

Ichthyoplankton data for each station are standardized to the number under 10 m^2 surface area. The Δ-distribution (Aitchison, 1955), is used to provide unbiased estimates of the sample means. Methods used to measure the effects of spatial and temporal differences in fish egg production on the precision estimates of total seasonal egg production derived from ichthyoplankton surveys are given in Pennington and Berrien (1984). The techniques, when applied to the mesoscale survey data for Atlantic mackerel, silver hake, and yellowtail flounder (<u>Limanda</u> <u>ferruginea</u>), produce estimates of total egg production with an average coefficient of variation equal to 31% that are sufficiently accurate to detect major population trends. Estimates of spawning stock size based on the

215

TABLE 10.1.

Average monthly chlorophyll for shoal Georges Bank 1977 through 1982.

Month	Average Chlorophyll (mg/m^3)	Coefficient of Variation (%)	Total Stations Sampled
Jan	-	-	-
Feb	-	-	-
Mar	6.20	35	26
Apr	-	-	-
May	2.24	64	34
Jun	2.53	38	12
Jul	1.83	38	26
Aug	1.86	46	12
Sep	1.83	41	10
Oct	2.71	24	19
Nov	2.36	50	20
Dec	1.73	16	15

Shallow Georges Bank < 60 m

TABLE 10.2.

Mean annual primary productivity comparison between 1940 Riley data and 1981 O'Reilly and Busch data for Georges Bank.

	$\bar{X}$ gCm^2/yr < 60 m	$\bar{X}$ gCm^2/yr 60-100 m	$\bar{X}$ gCm^2/yr > 100 m
Riley	530	360	290
O'Reilly & Busch	455	310	280

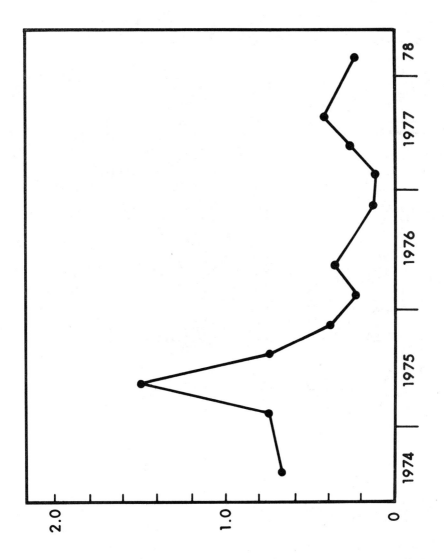

Figure 10.6. Cumulative carbon production of the Southern California Bight, 1974 through 1978. (From Lasker, 1978).

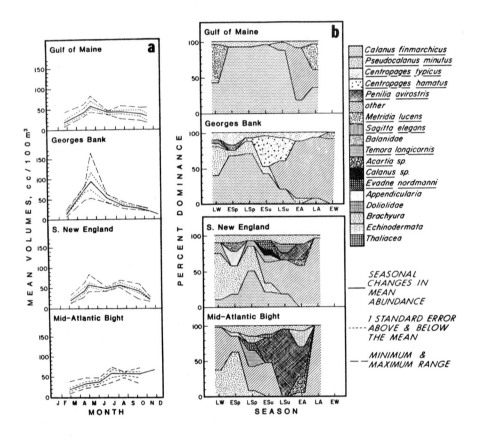

Figure 10.7.   Patterns of zooplankton in four
northeastern U.S. continental shelf subareas--Gulf of
Maine, Georges Bank, Southern New England, and the Mid-
Atlantic Bight.   (a) Seasonal patterns in mean
zooplankton standing stock (cc/100 m^3) for the 5-yr
MARMAP time series.   Solid line represents the mean,
short dashed line is one standard duration, and long
dashed line is the range.   (b) Seasonal patterns of
dominance of zooplankters by subarea shown as a
percentage of the samples with a dominant taxon in the
5-yr MARMAP time series.   LW = late winter, ESp = early
spring, ESu = early summer, EA = early autumn, LA =
late autumn, and EW = early winter.

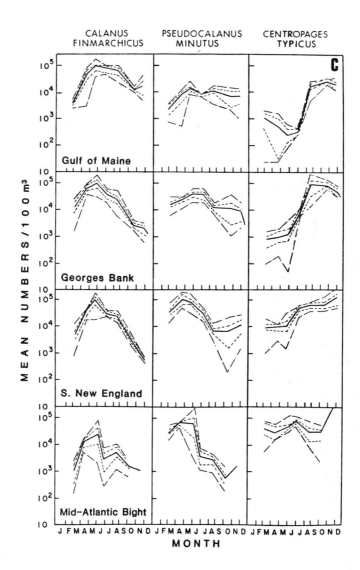

Figure 10.7 (continued). (c) Seasonal pulses in abundance of the three dominant copepod species Calanus
finmarchicus, Pseudocalanus minutus, and Centropages
typicus (No./100 m³) in each of the subareas for the 5-
yr time series. Solid line represents the mean, short
dashed line is one standard deviation, and long dashed
line is the range. (Adapted from Sherman et al.,
1983a.)

220

TABLE 10.3.

Summary of probability statistics for the two-tailed Fisher-sign test for year-to-year coherence in the zooplankton volumes, dominance, and three dominant species--C. finmarchicus, P. minutus, and C. typicus. Annual departures from the MARMAP 5-yr mean annual cycle were tested for each subarea. The ranges of the probability of the Fisher-sign statistic are tabulated. Of the 100 tests (5 yr x 4 areas x 5 variables) only four reject the null hypothesis at 0.05 significance. H_0: annual cycle = 5 yr mean cycle; * = significant difference in the year indicated in parentheses.

	Subarea			
Survey Variable	Gulf of Maine	Georges Bank	Southern New England	Mid-Atlantic Bight
MARMAP				
Volume	.344-.875	.031-.910	.227-.656	.109-.812
Dominance	.145-.773, .007* (78)	.109-.500	.172-.637	.188-.500
C. finmarchicus	.344-.773	.109-.891	.500-.656	.188-.812
P. minutus	.344-.773	.188-.891, .984* (79)	.344-.656, .992* (79)	.500-.812
C. typicus	.227-.891	.344-.891, .984* (77)	.227-.891	.500-.891

TABLE 10.4.

Comparisons of zooplankton volumes (cc/100 m^3) by subarea between MARMAP data and the earlier studies on the northeast continental shelf. No significant differences were found between MARMAP data and earlier studies in comparisons of displacement volumes (Kruskal Wallis P > 0.05) Volumes reported by Bigelow (1926) for late summer on Georges Bank were relatively high compared to those for the same season in MARMAP data. However, Bigelow's sampling was heavily biased towards the Northeast peak of Georges Bank. The range of mean displacement volumes for that region in the MARMAP data is 24.4 to 191.7 cc/100 m^3.

	Late Winter	Early Spring	Late Spring	Early Summer	Late Summer	Early Autumn	Late Autumn	Early Winter
Gulf of Maine								
MARMAP 1977-1981	10.9-47.0	34.6-65.2	44.0-83.2	40.3	31.8-58.0	23.3-57.5	18.4-53.9	
Bigelow, 1912-1920	17.8				25.5-47.7			
P	.380				.248			
Georges Bank								
MARMAP 1977-1981	11.4-24.0	50.2-86.5	56.2-166.0	46.2-65.8	31.4-43.9	25.8-37.2	23.2-28.8	13.9
Bigelow, 1912-1920	23.8				74.9			
P	.655				.157			
Southern New England								
MARMAP 1977-1981	13.2-33.5	32.0-66.5	46.7-85.4	43.4-54.4	57.4-69.2	24.2-60.9	21.4-28.4	
Bigelow & Sears, 1929-1932	8.7-19.5	59.6-72.3	42.5-93.0	40.3-89.3		38.0-40.6		
P	.180	.101	.631	.157		.770		
Grice & Hart, 1960	12			40	61	38	14	
P	.143			.180	.770	.380	.157	
Mid-Atlantic Bight								
MARMAP 1977-1981	11.8-39.6	25.2-51.5	29.5-50.9	41.0-73.2	50.4-66.0	37.4-76.0	70.1	
Bigelow & Sears, 1929-1932	33.6-39.1	27.0-48.7	24.7-75.1	38.6-52.4		44.8		
P	.180	.655	.715	.248		.380		

Seasons

egg production estimates compare favorably with other
independent assessments of stock size (Table 10.5).

A description of the methods used to estimate the
size of spawning biomass based on larval fish data, for
species not fished or sampled adequately by bottom
trawls is given in Berrien et al. (1984). Among those
species is the sand lance. From an analysis of
ichthyoplankton species composition and abundance data
we observed a population explosion of sand lance,
Ammodytes spp. from 1974 through 1981 (Morse, 1984),
coincident with a decline in Atlantic herring and
Atlantic mackerel (Figure 10.8). A similar coincident
shift in abundance occurred in the North Sea where the
declining herring and mackerel stocks appeared to be
replaced by increases in the populations of small,
fast-growing sprat, Clupea sprattus; sand lance, and
Norway pout Trisopterus esmarkii (Sherman et al.,
1981).

COMPARATIVE FOOD CHAIN VARIABILITY AND COHERENCE

The relative stability observed in phytoplankton
and zooplankton, when considered in relationship to the
decline in finfish biomass and subsequent population
explosion of the fast-growing, short-lived zooplanktivorous sand lance suggests that the reduction in
abundance of the mackerel, herring, gadoid, and
flatfish stocks were not attributable to a lack of food
at the lower end of the food chain. The conclusion of
Clark and Brown (1977) relating the finfish biomass
decline to excessive fishing mortality is supported by
the plankton record. It appears that fishing mortality
has imposed greater perturbations on fish populations
of the northeast shelf than any changes in plankton
abundance.

The plankton-fish linkage is apparently more
direct in other LMEs. In the California Current
Ecosystem, changes in primary production have been
correlated with recruitment in several stocks (Bakun
and Parrish, 1980). Cushing (1983) recently associated
the future of two year-classes of anchoveta off Peru to
a three-fold decline in primary production during the
El Niño of 1972-1973. Cushing (1982, 1983) further
suggests that the gadoid increase in the North Sea may
have been caused by the shift in dominance from small
copepods to larger copepods.

POTENTIAL FOR INCREASING YIELDS IN LMEs

The concept of managing multispecies fishery
resources within the spatial limits of a LME is
relatively recent. In 1972 ICNAF established the first
two-tiered system for managing the total finfish

223

TABLE 10.5.

Estimates of spawning stock size based on egg surveys and a cohort analysis.

Species	Estimate of Spawning Stock (egg surveys)	Confidence Interval	Estimate of Spawning Stock Size[a] (cohort analysis)
Atlantic mackerel	1.20×10^9	$(.31 \times 10^9, 2.12 \times 10^9)$*	$.96 \times 10^9$
Silver hake	1.55×10^9	$(1.14 \times 10^9, 1.96 \times 10^9)$*	$.77 \times 10^9$
Yellowtail flounder	1.38×10^8	$(.35 \times 10^8, 2.99 \times 10^8)$**	Not Available

*80% level.
**70% level.
[a]Resource Assessment Division, NMFS, NEFC, Woods Hole, MA.

224

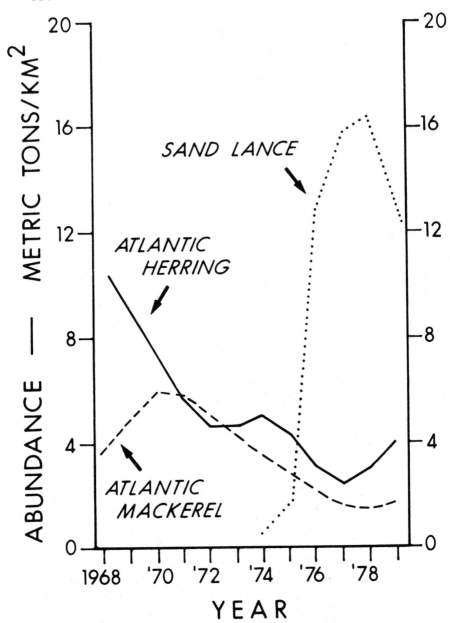

Figure 10.8. Decline of Atlantic herring and Atlantic mackerel and apparent replacement by the small, fast-growing sand lance in the northeast continental shelf ecosystem (measured in metric tons per sq km 1968-1979). (From Sherman et al., 1983b.)

biomass within the Northwest Atlantic ecosystem. Catches were capped at a level of one million metric tons annually for all species, and TACs were allocated for each of the important stocks comprising the biomass.

During the early 1970s the concept of fisheries ecosystem management for the North Sea was advocated by Andersen and Ursin (1978) of Denmark. Based on their fishery ecosystem models, they hypothesized that within the North Sea ecosystem predation and food concentration were density-dependent functions of both predator and prey abundance. They also hypothesized that the formerly abundant stocks of herring and mackerel presumably took major amounts of various kinds of fish larvae, and in turn herring and mackerel were preyed upon by large gadoids, particularly cod. Therefore it was concluded that the pelagic stocks would benefit from a reduction in the average size of gadoid predators. In fact the total yield of fish from the North Sea could be doubled based on their simulation results. That is an increase from 3 million tons to about 6 million tons, although the value of the catch would increase by a smaller amount because of the smaller size and greater abundance of the fish harvested.

A similar multispecies predator-prey model is under consideration at the Northeast Fisheries Center (NEFC) of NMFS (Sissenwine et al., 1984). The effects of predator-prey interactions and competition are being considered at NEFC based on the examination of 70,000 fish stomachs, including pelagic and demersal species in all stages of development. A total of 85% to 90% of the annual fish production on Georges Bank is consumed by predators including fish, and in lesser amounts by sea birds and marine mammals. Conceptually the total fish yield could be increased by a factor of two by selectively cropping the larger fish-eating species in the Georges Bank ecosystem. However, while this is conceptually possible, there remain significant uncertainties in the biological, economic, and technological interactions in a multispecies fishery that need to be carefully evaluated before encouraging selective predation on a large-scale.

FISHERY-INDEPENDENT ASSESSMENTS IN OTHER LMEs

In the Bering Sea and Gulf of Alaska both ichthyoplankton surveys and bottom trawl surveys are conducted to monitor changes in resources populations. It appears from preliminary analyses that both advective processes and predation interact to influence the size of year-classes of important species including the walleye pollock, Theragra chalcogramma; red king crab, Paralithodes camtschatica; and the Tanner crabs,

Chionoecete bairdi and C. opilio (Incze and Schumacher,
this volume).

In the Gulf of Mexico LME, bottom-trawl surveys
are conducted in the Mississippi Delta region to
monitor changes in fish stock abundance. In addition,
ichthyoplankton surveys over the entire Gulf of Mexico
initiated in 1977 are continuing. Station locations
are shown in Figure 10.9. At each location an ichthyoplankton tow is accompanied by water-column measurements of temperature, salinity, nutrients, chlorophyll,
and light penetration. In this high-diversity ecosystem (1,500 estimated species), studies are focused on
the taxonomy of fish eggs and larvae; estimates of the
spawning biomass of target species including bluefin
tuna, bluefish (Pomatomus saltatrix), and menhaden; and
the impact of pollutants in the Mississippi plume on
the growth and survival of larval menhaden, croaker,
and spot.

Joint studies among NMFS Fisheries Centers
involving trawling and acoustic surveys, ichthyoplankton surveys, and environmental observations are conducted on trans-ecosystem migrants including sharks,
tunas, billfish, and smaller species that have evolved
spawning strategies that utilize separate spawning
areas and nursery grounds characterized by Pacific hake
on the west coast (Figure 10.10) and bluefish on the
east coast (Figure 10.11).

FUTURE MANAGEMENT TRENDS IN LMEs

The combined multispecies assessment and environmental activities of the NMFS Fisheries Centers provide
a firm basis for measuring annual changes in target
populations and their environments throughout the EEZ.
Although fisheries interests are the principal users of
the information collected and synthesized by NMFS, the
opportunities for working across the different jurisdictions dealing with the management of entire ecosystems are growing.

There is an increasing public awareness of the
potential damage to marine ecosystems from the impacts
of ocean waste disposal of municipal sludges, radionuclides, petrogenic hydrocarbons, organochlorine
pesticides, and other hazardous substances. Within the
past decades, legislative authorities have been designated to deal with the regulation of offshore energy-
related development (DOE, DOI), protection of the marine environment (EPA), hazards to navigation (Corps of
Engineers), the protection of marine birds (Fish and
Wildlife Service), the protection of marine mammals
(Marine Mammal Commission), and yet other mandates,
which all underline the need for a more holistic
approach to the management of LMEs. Increasing use is

227

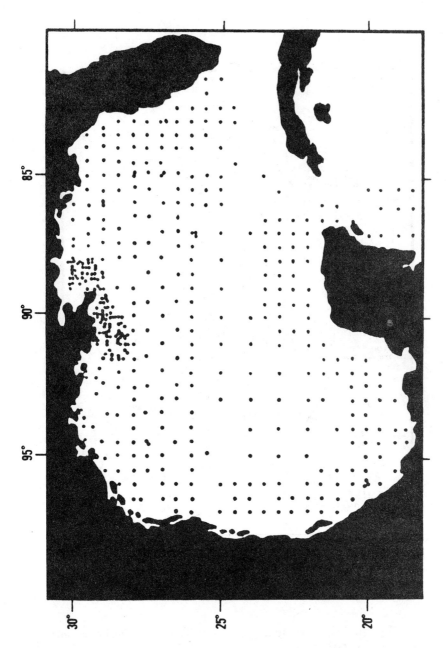

Figure 10.9.   Station pattern of the Gulf of Mexico
ichthyoplankton surveys, 1982.   (From Sherman et al.,
1983b.)

228

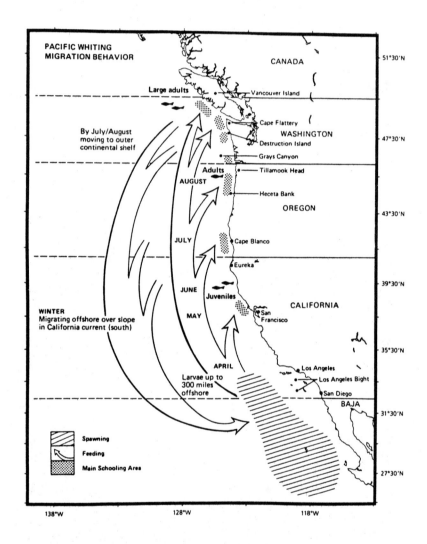

Figure 10.10. Migratory patterns of Pacific whiting
(hake) in relation to spawning, feeding, and schooling
areas in the California Current and Washington-Oregon
coastal ecosystems. (Adapted from Bailey et al.,
1982.)

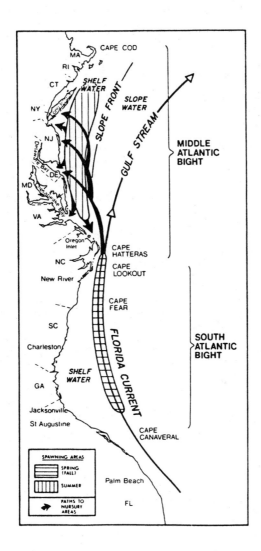

Figure 10.11. Spawning areas and coastal migration patterns for bluefish along the Atlantic coast. (From Kendall and Walford, 1979.)

made of Memoranda of Understanding to overcome institutional constraints within the federal government.
Significant interaction is growing among the federal
government, states, academia, and public sector for
dealing with overexploitation of marine resources.
However, as pointed out by MacCall (this volume),
significant jurisdictional barriers remain to be
overcome in order to achieve ecosystem management.

The international community is also moving toward
total ecosystem management. The most recent example of
the movement toward total ecosystem management is found
in the language of the Commission for the Conservation
of Antarctic Marine Living Resources. Article II of
the Convention calls for holistic management wherein
the regime should provide for the effective conservation of the living marine resources of the Antarctic
ecosystem as a whole. Considerable progress has been
made by the international scientific community in the
coordination and integration of studies in the
Antarctic leading to population assessments of the
principal ecosystem populations, including krill,
Euphausia spp. and its predators and prey under the
aegis of the Biological Investigations of Marine
Antarctic Systems and Stocks program (BIOMASS)
(Beddington and May, 1982). The approach in dealing
with the world's largest marine ecosystem has been a
combination of international krill biomass assessment
surveys, an analysis of catch data, and surveys of
marine birds, mammals, and fish (Laws, 1980; Everson,
1981; Pommeranz et al., 1981; Hureau, 1982; BIOMASS
Working Party on Fish Biology, 1982; BIOMASS Working
Party on Bird Ecology, 1982a,b,c).

For effective management of any LME it is necessary to survey the populations and their environments
(Brown et al., 1984). Unfortunately surveys can become
dull, routine affairs. However, they are critical
components of a total ecosystem resources assessment
program. Technical advances in hydroacoustics, satellite remote sensing of ocean features (Lasker et al.,
1981; Peláez and Guan, 1982), and electronic particle
sampling and data processing at sea (Lough and Potter,
1983) and in the laboratory (Jeffries et al., 1980),
when applied to the MARMAP type multispecies ichthyoplankton time-series surveys and target-species
recruitment studies will contribute to increased
sampling efficiencies and reduced costs of the
assessment surveys of large marine ecosystems.

NOTES

1. MARMAP Contribution No. MED/NEFC 84-13.
2. Mention of trade names or commercial firms
does not imply endorsement by the National Marine
Fisheries Service, NOAA.

REFERENCES

Ahlstrom, E. H. 1954. Distribution and abundance of egg and larval populations of the Pacific sardine. Fish. Bull., U.S. 56:82-140.

Ahlstrom, E. H. 1959. Vertical distribution of pelagic fish eggs and larvae off California and Baja California. Fish. Bull., U.S. 60:107-146.

Ahlstrom, E. H. 1965. Kinds and abundance of fishes in the California current region based on egg and larval surveys. State of California Marine Research Committee. Calif. Coop. Oceanic Fish. Rep. 10:31-52.

Ahlstrom, E. H. 1966. Distribution and abundance of sardine and anchovy larvae in the California Current region off California and Baja California, 1951-65: A summary. U.S. Fish Wildl. Serv. Spec. Sci. Rep.-Fish. 534, 71 p.

Ahlstrom, E. H., and Moser, H. G. 1981. Systematics and development of early life history stages of marine fishes: present status of the discipline and suggestions for the future. Rapp. P.-v. Réun., Cons. int. Explor. Mer 178:541-546.

Aitchison, J. 1955. On the distribution of a positive random variable having a discrete probability mass at the origin. J. Am. Statist. Assoc., 50:272, 901.

Andersen, K. P., and Ursin, E. 1977. A multispecies extension to the Beverton and Holt Theory of Fishing, with accounts of phosphorus circulation and primary production. Medd. fra Dan. Fisk. Havunders. NS 7:319-435.

Andersen, K. P., and Ursin, E. 1978. A multispecies analysis of the effects of variations of effort upon stock composition of eleven North Sea fish species. Rapp. P.-v. Réun. Cons. int. Explor. Mer 172:286-291.

Bailey, K. M., Francis, R. C., and Stevens, P. R. 1982. The life history and fishery of Pacific whiting, Merluccius productus. Calif. Coop. Oceanic Fish. Invest. Rep. 23:81-98.

Bakun, A., and Parrish, R. H. 1980. Environmental inputs to fishery population models for eastern boundary current regions. In Workshop on the effects of environmental variation on the survival of larval pelagic fishes, pp. 67-104. Ed. by G. D. Sharp. IOC Workshop Report 28, UNESCO, Paris.

Beddington, J. R., and May, R. M. 1982. The harvesting of interacting species in a natural ecosystem. The changing populations of whales and other animals that feed on the krill of the Southern Ocean are an example of the problem of utilizing a biological resource without extinguishing species. Sci. Am. 247(5):62-69.

Beddington, J. R., May, R. M., Clark, C. W., Holt, S. J., and Laws, R. M. 1979. Management of multispecies fisheries. Science 204(4403):267-277.

Berrien, P., Morse, W., and Pennington, M. 1984. Recent estimates of adult spawning stock biomass for important fish stocks off northeastern United States from MARMAP ichthyoplankton surveys. U.S. Dep. Commer., NOAA Tech. Mem. NMFS F/NEC-30. 111 pp.

BIOMASS Working Party on Bird Ecology. 1982a. Recording observations of birds at sea. BIOMASS Handbook 18. SCAR/SCOR/IABO/ACMRR Group of Specialists on Living Resources of the Southern Ocean.

BIOMASS Working Party on Bird Ecology. 1982b. Monitoring studies of seabirds. BIOMASS Handbook 19. SCAR/SCOR/IABO/ACMRR Group of Specialists on Living Resources of the Southern Ocean.

BIOMASS Working Party on Bird Ecology. 1982c. Penguin census methods. BIOMASS Handbook 20. SCAR/SCOR/IABO/ACMRR Group of Specialists on Living Resources of the Southern Ocean.

BIOMASS Working Party on Fish Biology. 1982. Recommended methods for standardization of measurements of fish. BIOMASS Handbook 13. SCAR/SCOR/IABO/ACMRR Group of Specialists on Living Resources of the Southern Ocean. D. Sahrhage, Chairman.

Brown, B. E., Anthony, V. C., Anderson, E. D., Hennemuth, R. C., and Sherman, K. 1984. The dynamics of pelagic fishery resources off the northeastern coast of the United States under conditions of extreme fishing perturbations. North West Atlantic--Atlantico Nordoccidental. FAO Fish. Rep. 2(291):465-505.

Clark, S. H., and B. E. Brown. 1977. Changes of biomass of finfishes and squids from the Gulf of Maine to Cape Hatteras, 1963-74, as determined from research vessel survey data. Fish. Bull., U.S. 75:1-21.

Colebrook, J. M. 1977. Annual fluctuations in biomass of taxonomic groups of zooplankton in the California Current, 1955-59. Fish. Bull., U.S. 75:357-368.

Colebrook, J. M. 1978a. Changes in the zooplankton in the North Sea, 1948-1973. Rapp. P.-v. Réun. Cons. int. Explor. Mer 172:390-396.

Colebrook, J. M. 1978b. Continuous plankton records: zooplankton and environment, Northeast Atlantic and North Sea, 1948-1975. Oceanologia Acta 1(1):9-23.

Cushing, D. H. 1982. Sources of variability in the North Sea ecosystem. Meeting on the North Sea. University of Hamburg, 3-8 September, 1981.

Cushing, D. H. 1983. The outlook for fisheries research in the next ten years, In Global fisheries perspectives for the 1980s. pp. 263-277. Ed. by B. J. Rothschild. Springer-Verlag New York Inc., N. Y.

Evans, C., and O'Reilly, J. P. 1983. A manual for the measurement of Chlorophyll a, net phytoplankton and nannoplankton. BIOMASS Handbook 9. SCAR/SCOR/IABO/ACMRR Group of Specialists on Living Resources of the Southern Ocean.

Everson, I. 1981. Antarctic fish age determination methods. BIOMASS Handbook 8. SCAR/SCOR/IABO/ACMRR Group of Specialists on Living Resources of the Southern Ocean.

Grosslein, M. D., Langton, R. W., and Sissenwine, M. P. 1980. Recent fluctuations in pelagic fish stocks in the Northwest Atlantic, Georges Bank region, in relation to species interactions. Rapp. P.-v. Réun. Cons. int. Explor. Mer 177:374-404.

Hureau, J-C. 1982. Methods for studying early life history stages of Antarctic fishes. BIOMASS Handbook 17. SCAR/SCOR/IABO/ACMRR Group of Specialists on Living Resources of the Southern Ocean.

Incze, L. S., and Schumacher, J. D. This volume. Influence of mesoscale environmental events on upper trophic level biomass of the Eastern Bering Sea.

Jeffries, H. P., Sherman, K., Maurer, R., and Katsinis, C. 1980. Computer-processing of zooplankton samples. In Estuarine perspectives, pp. 303-316. Ed. by V. S. Kennedy Acad. Press, Inc., N.Y.

Jones, R. 1976. An energy budget for North Sea fish species and its application for fish management. ICES C.M. 1976/F:36.

Jossi, J. W., and Marak, R. R. 1983. MARMAP plankton survey survey manual. U.S. Dep. Commer., NOAA Tech. Mem. NMFS F/NEC-21. 258 pp.

Kendall, A. W., Jr., and Walford, L. A. 1979. Sources and distribution of bluefish, Pomatomus saltatrix, larvae and juveniles off the east coast of the United States. Fish. Bull., U.S. 77:213-227.

Kramer, D., and Smith, P. E. 1971. Seasonal and geographic characteristics of fishery resources: California Current Region--V. Northern anchovy. Commer. Fish. Rev. 33(3):33-38.

Laevastu, T., and Favorite, F. 1978. Numerical evaluation of marine ecosystems. Part I. Deterministic bulk biomass model (BBM). NMFS

Northwest and Alaska Fisheries Center, Seattle, Wash., Processed Report. 22 pp.

Laevastu, T., and Favorite, F. 1981. Holistic simulation of marine ecosystem. In Analysis of marine ecosystems, pp. 702-727. Ed. by A. R. Longhurst. Acad. Press, Inc., Lond.

Laevastu, T., and Larkins, H. A. 1981. Marine fisheries ecosystem: Its quantitative evaluation and management. Fishing News Books Ltd., Farnham, Surrey, Engl.

Lasker, R. 1978. The relation between oceanographic conditions and larval anchovy food in the California Current: Identification of factors contributing to recruitment failure. Rapp. P.-v. Réun. Cons. int. Explor. Mer 173:212-230.

Lasker, R. 1981a. Factors contributing to variable recruitment of the northern anchovy (Engraulis mordax) in the California Current: Contrasting years, 1975 through 1978. Rapp. P.-v. Réun. Cons. int. Explor. Mer 178:375-388.

Lasker, R. (Editor). 1981b. Marine fish larvae-- morphology, ecology, and relation to fisheries. Washington Sea Grant Program, Univ. Washington Press, Seattle and London.

Lasker, R., Pelaez, J., and Laurs, R. M. 1981. The use of satellite infrared imagery for describing ocean processes in relation to spawning of the northern anchovy (Engraulis mordax). Remote Sens. Environ. 11:439-453.

Laws, R. M. 1980. Estimation of population sizes of seals. SCAR/SCOR/IABO/ACMRR Group of specialists on Living Resources of the Southern Ocean. BIOMASS Handbook 2.

Lough, R. G., and Laurence, G. C. 1982. Larval haddock and cod survival studies on Georges Bank. In Report of the Working Group on Larval Fish Ecology. Lowestoft, England 3-6 July 1981. pp. 103-119. ICES C.M.1982/L:3.

Lough, R. G., and Potter, D. C. 1983. Rapid shipboard identification and enumeration of zooplankton samples. J. Plankton Res. 5:775-782.

MacCall, A. D. This volume. Changes in the biomass of the California Current ecosystem.

Mann, K. H. 1982. Ecology of coastal waters: A systems approach. Studies in ecology vol. 8. Univ. of Calif. Press, Berkeley. 322 pp.

Morse, W. 1984. An assessment of the Georges Bank haddock stock based on larvae collected on MARMAP plankton surveys, 1977-1982. In Recent estimates of adult spawning stock biomass for important fish stocks off northeastern United States from MARMAP ichthyoplankton surveys. U.S. Dep. Commer., NOAA Tech. Mem. NMFS-F/NEC-30. 111 pp.

O'Reilly, J., and Busch, D. 1984. Phytoplankton primary production for the Northwestern Atlantic shelf. Rapp. P.-v. Réun. Cons. int. Explor. Mer 183:255-268.

O'Reilly, J., and Thomas, J. P. 1983. A manual for the measurement of total daily primary productivity. SCAR/SCOR/IABO/ACMRR Group of Specialists on Living Resources of the Southern Ocean. BIOMASS Handbook 10.

Parrish, J. D. 1975. Marine trophic interactions by dynamic simulation of fish species. Fish. Bull., U. S. 73:695-716.

Peláez, J., and Guan, F. 1982. California Current chlorophyll measurements from satellite data. Calif. Coop. Oceanic Fish. Invest. Rep. 23:212-225.

Pennington, M., and Berrien, P. 1982. Measuring the effect of the variability of egg densities over space and time on egg abundance estimates. In Report of the Working Group on Larval Fish Ecology. Lowestoft, England 3-6 July 1981. pp. 127-141. ICES C.M.1982/L:3.

Pennington, M., and Berrien, P. 1984. Measuring the precision of estimates of total egg production based on plankton surveys. J. Plankton Res. 869-879.

Pommeranz, T., Herman, C., and Kuhn, A. 1981. Data requirements of estimating krill abundance using standard net sampling equipment. SCAR/SCOR/IABO/ ACMRR Group of Specialists on Living Resources of the Southern Ocean. BIOMASS Handbook 12.

Posgay, J. A., and Marak, R. R. 1981. The MARMAP bongo zooplankton samplers. J. Northw. Atl. Fish. Sci. 1:91-99.

Regier, H. A., and Henderson, H. F. 1973. Towards a broad ecological model of fish communities and fisheries. Trans. Am. Fish. Soc. 102:56-72.

Sheldon, R. W., Sutcliffe, W. H. Jr., and Paranjape, M. A. 1977. Structure of pelagic food chain and relationship between plankton and fish production. J. Fish. Res. Board Can. 34:2344-2353.

Sherman, K., Cohen, E., Sissenwine, M., Grosslein, M., Langton, R., and Green, J. 1978. Food requirements of fish stocks of the Gulf of Maine, Georges Bank, and adjacent waters. ICES C.M.1978/Gen:8 (Symposium).

Sherman, K., Green, J. R., Goulet, J. R., and Ejsymont, L. 1983a. Coherence in zooplankton of a large northwest Atlantic ecosystem. Fish. Bull., U.S. 81:855-862.

Sherman, K., Jones, C., Sullivan, L., Smith, W., Berrien, P., and Ejsymont, L. 1981. Congruent shifts in sand eel abundance in western and

236

eastern North Atlantic ecosystems. Nature 291(5815):486-489.

Sherman, K., Lasker, R., Richards, W., and Kendall, A. W., Jr. 1983b. Ichthyoplankton and fish recruitment studies in large marine ecosystems. Mar. Fish. Rev. 45(10-11-12):1-25.

Sissenwine, M. P., Cohen, E. B., and Grosslein, M. D. 1984. Structure of the Georges Bank ecosystem. Rapp. P.-v. Réun. Cons. int. Explor. Mer 183:243-254.

Smith, P. E., and Richardson, S. L. 1977. Standard techniques for pelagic fish egg and larva surveys. FAO Fish. Tech. Paper 175, Rome.

Stauffer, G. D., and Charter, R. L. 1982. The northern anchovy spawning biomass for the 1981-82 California fishing season. Calif. Coop. Oceanic Fish. Invest. Rep. 23:15-19.

TRW Systems Group. 1973a. MARMAP system description. TRW Systems Group, Redondo Beach, California. MARMAP Program Office, Wash., D.C. January 1973. NTIS No. COM-74-10829.

TRW Systems Group. 1973b. Survey 1 plan for MARMAP. TRW Systems Group, Redondo Beach, California. MARMAP Program Office, Wash., D.C. January 1973. NTIS No. COM-74-10827.

TRW Systems Group. 1974. Survey 2 plan for MARMAP. TRW Systems Group, Redondo Beach, California. MARMAP Program Office, Wash., D.C. October 1974. NTIS No. PB80-113343.

Wickett, W. P. 1967. Ekman transport and zooplankton concentration in the North Pacific Ocean. J. Fish. Res. Board Can. 24:581-594.

Part 3

Institutional Framework
for Managing
Large Marine Ecosystems

11. Introduction to Part Three: Large Marine Ecosystems as Regional Phenomena

The designation of large marine ecosystems (LMEs) as functional units is an exercise in applied region- alism, and is therefore of interest to geographers as well as to fishery scientists, fishery economists, and lawyers. A region may be defined as an area of the earth's surface within which there exists a particular feature or association of features. The region should be delimited by boundaries, however broadly defined, in order to separate it from neighboring areas in which the distinguishing characteristics are not present, or exist only in a markedly diluted form. A region may vary in size from macro- to micro-geographic, and there should be a sufficient variety of features present to distinguish it from a simple isolated phenomenon, such as an oil spill.

A region is essentially an intellectual concept, and while designated regions should be based on the existence of observable phenomena, it is up to the com- piler of the region's description to define in detail the nature and extent of the characteristic feature or features, the forms of interaction they may have with one another and with other aspects of the environment, the boundaries of the region, the existence or non- existence of subregions, and the presence of other regions of similar scale and intensity in other parts of the world. The region should be capable of being represented on a map or chart, and conditions within the region should be measurable.

In this volume we encounter a variety of LMEs, located in various regions of the world, at differing scales of magnitude. Seven LMEs were identified within parts of the U.S. Exclusive Economic Zone: Insular Pacific, Eastern Bering Sea, Gulf of Alaska, California Current, Gulf of Mexico, Southeast Atlantic Shelf, and Northeast Atlantic Shelf. We heard of the Scotian Shelf and the Eastern Canadian Arctic, as well as Georges Bank (a subregion of the Northeast Atlantic Shelf), the easternmost part of which lies beyond the limits of the U.S. EEZ. Other LMEs we have been

informed about include the North and Baltic Seas, the
Eastern Tropical Pacific, the Peruvian coastal waters,
and, last but not least, the waters surrounding
Antarctica--an area which some of us hope will eventually be referred to as the "Southern Ocean."

What qualities do these LMEs seem to have in common? I would mention, first, that the state of scientific investigation of LMEs is still in an early stage.
For one thing, the systems discussed in this volume are
by no means inclusive, and some years may pass before
we can expect a global "Atlas of Large Marine Ecosystems." Another indication of the relatively early
stage of investigation is the apparent difficulty
encountered in defining subregions (or perhaps subsys-
tems) within the areas now under study. For example, I
would suggest that in time subcategories of the Antarctic Living Marine Ecosystem may come to be identified.

A second common theme, and one that is addressed
in this session, is the need for greater analysis of
the legal, economic, and other institutional parameters
of these large ecosystems. The 1982 Law of the Sea
Convention practically ignored the transnational nature
of many LMEs, except to urge (Article 123) that States
bordering an enclosed or semi-enclosed area should
endeavor "to co-ordinate the management, conservation,
exploration and exploitation of the living resources."
But some LMEs are not located within semienclosed seas.
Further, as Belsky points out, there has for some time
been talk of "total ecosystem management" of living
marine resources--a process which would involve far
greater certainty about both the physical and the
institutional aspects of LMEs than now exist.

Finally, since we have developed a new creature,
the LME, we need to seriously consider optimum methods
for regulating its exploitation. As both Christy and
Pontecorvo point out in their papers, the traditional
approach to fisheries management has been on a species-
by-species basis. Processes could be measured and
regulations devised for achieving desired objectives
from the fishery. But how does one measure conditions
in a multispecies system? How does one accommodate the
separate objectives of separate harvests, both by one
country's fishermen, and at times by several countries'
fishermen? What new hypotheses and new data management
techniques must be developed? If nothing else, this
section on Institutional Frameworks points up the need
for increased input into the study of evolving method-
ologies of LME management by lawyers, economists,
political scientists, and representatives of the fishing industry complex. Otherwise, the early identification and assessment of critical problems in the
management process of this new phenomenon, the LME,
will be very difficult to achieve.

12. Legal Constraints and Options for Total Ecosystem Management of Large Marine Ecosystems

ABSTRACT

One reason for the continuing threats to coastal and marine ecosystems is the historical legal framework for regulating activities in the oceans. This chapter will describe that legal superstructure and how it has been evolving in the last twenty years to provide in- creased opportunities for total ecosystem management of the oceans. It will conclude with suggestions on how individuals and governments can promote total ecosystem management and secure its establishment as a basic tenet of ocean law and policy.

INTRODUCTION

The Legal Framework

The international scientific community has long recognized that the oceans consist of interrelated species and that actions that affect any part of a marine ecosystem will necessarily affect the rest of that system (Ross, 1982). This has led to calls for total ecosystem management (Gordon, 1981). To be effective, such total ecosystem management requires procedures and standards for the conservation and exploitation of living marine resources, for the study and protection of those resources and their habitats and for consideration of the whole system encompassing the resources and habitats (Springer, 1983). The international law of the sea, however, does not deal with total ecosystems. It focuses on the par- ticular activities of the citizens of particular states. Thus, at least until very recently, interna- tional law established few requirements for protection and management of the ocean ecology. Nation-states could, but did not have to, establish standards for their own nationals and their own waters. As to activities not within the explicit competence of a state, there were few, if any, rules. Any rules that existed were the result of voluntary agreements among states (Churchill and Lowe, 1983).

Even when states did act, either individually or collectively, to establish rules for ocean management, they separated out the questions of pollution control and resource exploitation. Domestic legislation and multilateral agreements only addressed particular issues such as pollution from vessels or coastal activities, or fishing for certain species in designated waters (Chapman 1970; Lutz, 1976; Waldichuck, 1982). Ocean management rules, whether established by statute or treaty, did not consider the relationships of species in and to an ecosystem (Comment, 1977).

THE PROBLEM OF JURISDICTION

Unlike domestic law established by legislatures or the political executives of nation-states, international rules are ordinarily not promulgated by any world-wide legislature or agency but rather are established through the customs and practices of nations or in multilateral compacts voluntarily entered into by sovereign and independent countries. When international laws are broken, enforcement is ordinarily the responsibility of each nation, either individually, or when nations have agreed, collectively through multinational institutions (Art. 38(1) Statute of the International Court of Justice; Brierley, 1963). Because of this reliance on cooperative action, most international prescriptions are limited in scope.

These general jurisprudential concepts govern the legal framework for activities in the coastal and ocean waters. National sovereignty is the primary basis for legal rules and for institutions to enforce these rules. Nation-states are generally allowed to decide what and how activities are to be conducted in their own territory. Both inside and outside of their territory, countries also have the power to establish rules for their own nationals (Wildenhus's Case, 1887; U.S. v. Flores, 1933; Lauritzen v. Larson, 1953). While some minimal limitations on such power have been established by custom and practice, restrictions on these sovereign rights ordinarily must be voluntarily agreed to by the nation-state and enforced by the nation-state alone or together with other nations pursuant to voluntary compacts or agreements (Saudi Arabia v. Arabian American Oil Co., 1958). Thus, nation-states are free to promulgate whatever rules they wish for any ecosystem within their territory. But they are also free, subject to minimal limitations found in customary international law or in voluntary compacts with other nations, not to establish any rules for such management.

When an ecosystem crosses over the territory of more than one nation, management of the total ecosystem can only occur with the explicit consent of each nation-state involved. Thus, each state can set or not set rules for management within its territory and completely disregard the standards established or not established by its sister state or states (Springer, 1983).

As difficult as these jurisdictional problems are in establishing rules for total ecosystem management in one or more nations' territorial waters, they are even more troublesome in dealing with ecosystems solely or partially within international waters. Until recently, international law assumed that international waters were res nullius, that is, totally free and belonging to no one and no nation. Except for controls a country placed on its own nationals, no rules or procedures could be established to preclude or even control activities in such waters. Nation-states were free to join together to establish common rules binding on their own people, but the citizens of other countries, not part of such treaty arrangements, were not bound by such standards. As with practices within territorial waters, some minimal limitations could be set by international custom and practice, but again these rules existed without effective international legal enforcement. Application of any common set of rules depended on voluntary acquiescence or the political power of other nations, either individually or acting through collective institutions (Goldberg et al., 1975; de Klemm, 1981).

With ecosystems existing within one or more nations' territory and in international waters, the jurisdictional responsibilities and limitations clash. Nation-states were free to establish their own rules, or preclude any rules, applicable to those parts of the ecosystem within their own waters. Even when states agreed to establish common standards, these restrictions only applied to the common territory and only to the nationals of the agreeing states in international waters. Unless there was a treaty, or a customary rule that all states were willing to follow, no state was forced to comply with these standards in dealing with those parts of the ecosystem that were in international waters. Moreover, because of nationalism, many states opposed establishment of such international standards that may have seemed to restrict their freedom to regulate activities within their territory or in international waters (Chapman, 1967; de Klemm, 1981).

The effect of these axioms of international law is that management of large marine ecosystems depends

almost entirely on the voluntary agreement of individual countries. To establish any standards, nations
must believe that comprehensive management is in their
own best interest. There are few legal impediments to
action; neither are there many legal incentives. This
provides opportunities as well as constraints. The
larger the territorial area within the jurisdiction of
a nation, the more scope there is for a sensitive
national political leadership to develop comprehensive
management programs (National Advisory Committee on
Oceans and Atmosphere, 1982). The greater the consensus there is by the international community as to the
need for comprehensive management, the more likely it
is that bilateral or multilateral standards can be
adopted and implemented to assure total ecosystem
management (Goldie, 1975). However, national governments have not focused on ecosystem management and
there is no rule of international law that forces them
to take a broad look at the whole set of problems that
face marine ecosystems. Rather, both state practice
and multi-state agreements have taken a more narrow
view of ocean management. Limits on pollution and
management of living resources have historically been
addressed separately.

CONTROLS ON POLLUTION AND FISHING

As a result of increased environmental sensitivity
during the last decade and a half, more and more
nations are adopting stringent laws and regulations to
control pollution of the coastal and ocean waters
adjacent to their coasts (Bender, 1981; Chasis,
1981). Such rules include limits on dumping of wastes
into the oceans and the reduction of land-based discharges into the oceans (Whipple, 1982). In addition,
countries have established rules to control the activities of their nationals in territorial or high seas.
Such rules have set design standards for vessels under
a nation's flag, for example, and have limited oil or
chemical discharges from those vessels (Schneider,
1982; Wallace and Ratcliffe, 1983). Finally, some
states have established rules for any vessel coming
into their ports. These rules are intended to assure
the safety of such vessels and this also results in
reducing the potential for pollution by such vessels
when in the waters of another state or in the high seas
(Ports and Waterways Safety Act, 1972).
In addition, transnational pollution is being
addressed by the international community. As early as
the 1950s, nation-states joined together in voluntary
bilateral and multilateral agreements to set minimum
standards and thus begin to limit the adverse environmental effects of man's activities. These agreements,

however, focused on particular subjects. Thus, separate treaties provided for the control, cleanup, and liability for oil spills of tankers, for limitations on the dumping of matter into the oceans, and for the minimization of pollution in connection with exploitation of seabed resources (e.g., Art. 5(7), Convention on the Continental Shelf, 1958; Art. 24, Convention on the High Seas, 1958; the International Convention on Civil Liability for Oil Pollution Damage, 1970; the Convention on the Prevention of Marine Pollution by Dumping of Wastes and Other Matter, 1972; the International Convention for the Prevention of Pollution from Ships, 1973). Moreover, these agreements were the result of political compromises and thus often minimal or general in scope. Even environmentally sensitive nations were concerned about their national sovereignty, and the international legal and political implications of setting too stringent international standards to control the activities in a national's territory or over a nation's citizens. They were even more reluctant to establish any international enforcement machinery to enforce these standards (Friedheim, 1975; Remond-Gouilloud, 1981b; Springer; 1983).

More recently, nations have joined together to develop standards to control pollution for particular geographic regions, where jurisdictions overlap, such as in the Baltic, the Mediterranean, and North Sea. Such "regional seas" programs and controls are more comprehensive in dealing with all types of pollution and more stringent in setting standards and in establishing procedures to enforce them (de Klemm, 1981; Johnston and Enomoto, 1981; Keckes, 1981; Boxer, 1983).

Nation-states are also acting to control overexploitation of living marine resources. Many states have established fisheries management procedures for fishing in their waters (e.g., Magnuson Fisheries Conservation and Management Act, 1976), and have joined together with adjacent states in agreements to govern fishing in waters that cross jurisdictional boundaries (Churchill and Lowe, 1983). Generally, however, such agreements, as with those on pollution, deal with particular narrow problems. Thus, they focus on particular species, on particular regions, or on protection of unique and sensitive endangered species or marine mammals (Travalio and Clement, 1979; Wilkinson and Conner, 1983). These agreements do not look to the whole ecosystem and the relationship between species. Moreover, even when they address geographic areas, these agreements usually focus on economic and political issues, such as dividing the resource, rather than on ecological issues, such as conservation of the resources (Farnell, 1981; Larson, 1983).

The Evolving Law of the Sea

As shown in the previous discussion, pollution and
living resource management issues have been addressed
by law, but not comprehensively. Still, international
law does not preclude total ecosystem management, and
there is some evidence that two recent developments in
international law can lead to future rules providing
for such management. The establishment of extensive
fisheries zones extending territorial responsibility as
far out as two hundred miles from a country's coast
provides the opportunity for states to individually
manage larger ecosystems and thus could force states to
re-evaluate their fisheries management policies.
Secondly, the international community is more willing
to accept an international responsibility for preserving resources in, and the ecology of, both territorial and international waters (Goldie, 1975;
Hargrove, 1975; Springer, 1983). This acceptance may
be merging into customary rules of international law
that promote consideration of total ecosystems and the
establishment of standards for those systems.

FISHERIES CONSERVATION AND MANAGEMENT ZONES

The extent of a nation's power over resources in,
and the ecosystem of, waters adjacent to its coast has
historically focused on the balance of the two international law doctrines described in the previous
section--freedom of the seas and adjacent state
sovereignty. Thus, a coastal state had territorial
dominion only over a very narrow band of neighboring
coastal waters (generally a three mile zone). Outside
that limited band, the citizens of all nations had the
right to use the oceans for any innocent purpose,
defined to include navigation and fishing. As
described earlier, except for controls over a country's
own nationals, no country had the unilateral authority,
let alone responsibility, to control such innocent
activities in such international waters.

The application of these rules mean few jurisdictional conflicts but also few standards. With
narrow three-mile zones and generally-accepted procedures to define these areas, few disputes existed as to
the power of a state to decide what could be done in
these territorial waters. Moreover, because of the
limited breadth of sovereign rights, most countries did
not establish rules for management or protection.
Other activities that might pollute the waters or
otherwise affect the habitat of living marine resources
were similarly not restricted (Magnuson, 1977).

The narrowness of sovereign claims meant a large
ocean area with almost no controls or restrictions.

The freedom to fish, or overfish, was an essential element of the traditional legal doctrine of freedom of the seas and not to be limited (d'Amato and Hargrove, 1975; de Klemm 1981). Similarly, the problem of the adverse effect on resources caused by conflicting uses of the oceans was not considered. Marine pollution was not perceived as a significant problem (Burke, 1967; Remond-Gouilloud, 1981a,c; Schneider, 1981).

It was not until 1945 that nations recognized the economic potential of the oceans and the possibility of reserving large amounts of that potential for their own citizens. In that year, President Truman issued a proclamation that asserted United States' jurisdiction over all the mineral resouces in the lands beneath the oceans out to the end of America's continental shelf (Proclamation No. 2667, 1945). While this Proclamation became the focus of international legal attention, the President issued another proclamation at the same time. In the second one, he indicated the possibility that the United States could limit and manage fishing off its coast through the creation of extended fishing conservation zones adjacent to the American coast (Proclamation No. 2668, 1945). The hope was that a warning of the possibility of future limitation could lead to reductions in foreign overfishing. It was not the United States' desire to create such zones then, as its fishing fleet was considered one of the best and it depended on the freedom to fish all over the globe (Jacobson and Davis, 1983).

Still, the precedent was set. In the 1950s and 1960s, Latin American countries claimed jurisdiction out to 200 miles so as to control all fishing in their waters and, in fact, attempted to exercise their claim of jurisdiction over the vigorous protests of the United States and other maritime nations (Hollick, 1981). The larger maritime nations objected to such zones as being contrary to international customary law, which provided for only a narrow territorial sea and for high seas freedoms beyond that area. Only treaty agreements, it was urged, could provide for such a radical change in the balance set between territorial sovereignty and international freedom of the seas (Schaefer, 1970; Knight, 1975).

Despite these protests, the concept of an extended fisheries zone began to evolve through discussions in diplomatic conferences in the late 1960s and early 1970s (Schaefer, 1970; Comment, 1983). More and more nations adopted or accepted such zones (Alexander, 1983). Finally, in the mid-1970s, the leading maritime nation, the United States, established its own 200 mile fishery conservation and management zone. In its 1976 Fisheries Conservation and Management Act, the United States claimed exclusive national sovereignty and control over all living marine resources out to 200 miles. The statute did not explicitly provide for rules and standards for ecosystems within such zones. Rather, it called for policies and regulations to govern and limit fishing of species (Greenberg and Shapiro, 1982). The statute does allow ecosystem management, however. Regional councils and federal regulators are free to develop controls for the conservation and management of groups of species and their habitats (Gordon, 1981). The statute only limited the scope of U.S. jurisdiction as to one category of fisheries--it precluded national regulation of highly migratory species (Larson, 1983).

The action by the United States confirmed the trend towards acceptance of 200 mile fishing zones as part of international customary law. In the 1980s, the concept of 200 mile fishing zones has become law. The proposed Law of the Sea Treaty explictly grants the legal authority to a nation to manage and control fisheries, as part of its sovereign power, in a 200 mile exclusive economic zone adjacent to its coast (Arts. 55 to 75, UNCLOS, 1982). Even those nation's which do not accept the Treaty, such as the United States, do accept the international legal existence of such 200 mile zones for all coastal nations (Proclamation No. 5030, 1983; Malone, 1984).

The existence of such zones presents both constraints and opportunities for total ecosystem management. With extended jurisdiction, it is more likely that an ecosystem, or at least large parts of an ecosystem, are within one country's territorial waters (de Klemm, 1981; Burke, 1983). Thus, an enlightened state can design comprehensive management regimes to deal with an ecosystem. Such management regimes can obviously include restrictions on fishing and rules and regulations to conserve and manage fisheries. They, however, can also include consideration of the whole ecological mosaic, involving the relationship between species and controls to limit adverse environmental effects on the affected ocean waters including the habitat of a fishery (Churchill and Lowe, 1983).

The primary constraint caused by such extended zones is the increased possibility of jurisdictional conflicts. Two-hundred mile national claims of sovereignty will mean more states claiming exclusive authority over activities in contested areas. Similarly, the likelihood of an ecosystem crossing over territorial boundaries is increased. Fewer systems will be exclusively in international waters. Even this constraint, however, can lead to opportunities. Countries will have to resolve disputes over territorial claims that did not heretofore exist. The need for resolution of those disputes could lead enlightened

leaders to realize the value of considering the whole
ecosystem in any bilateral or multilateral agreement
that would affect not just the contested area but all
of the fisheries zones of each affected nation-state
(de Klemm, 1981; Kildow, 1982).

While the potential opportunity for ecosystem
management is present, it must be noted that this is
only an opportunity and nothing more. The expansion of
nation-states' sovereignty over fisheries, habitats,
and ecosystems does not assure total ecosystem management regimes. The decision to provide for such comprehensive consideration is still a political one. Moreover, even if such a political decision is made by adjacent nation-states, it does not affect the resources
and activities beyond each country's exclusive economic
zone. For resources, habitats, and ecosystems
partially or totally within international waters, only
the international community can set and enforce rules
and standards. Fortunately, another international
legal trend provides opportunities for encouraging
enlightened nation-state action as to ecosystems within
their own waters, for promoting multi-state cooperation
for ecosystems in shared waters, and for securing
international concerted action for ecosystems partially
or totally within international waters.

THE OCEANS AS GLOBAL COMMONS

As noted earlier, traditional international law
doctrine divided the oceans into two jurisdictional
categories--those waters within the territorial
sovereignty of an adjacent coastal state and those
waters included in the open seas. No other nation or
group of nations could tell an adjacent state how to
manage activities within the waters under the state's
sovereignty. Similarly, no nation or group of nations
could limit the activities of anyone but its own
nationals on the high seas. Only voluntary agreements
or generally accepted rules established by custom and
practice could limit these freedoms. As a result,
until the middle of the twentieth century, there were
few international standards relating to ocean use and
abuse. Only recently have more and more nations and,
in fact, the organized international community, become
sensitive to the unique nature of the oceans. Individual nations have adopted more stringent environmental
rules and have joined with other nations in agreements
to provide for study of the ocean ecology, set minimum
standards to control pollution, establish a common
legal framework to resolve jurisdictional disputes, and
even to impose limited responsibility for fisheries
protection and exploitation (Alexander, 1967; Schaefer,
1970; Sohn, 1973; McManus, 1982; Burke, 1983).

Through these state practices and multilateral agreements, international oceans law has been changing. Most scholars, and many political leaders, believe that customary international law now provides that the oceans are the unique responsibility of the world community and that all nation-states share the obligation to assure its continued suvival as an international resource (d'Amato and Hargrove, 1975; Statement by President Reagan Accompanying Proclamation No. 5030, 1983). Nations must take appropriate steps to protect the ocean areas under their sovereignty and work with each other to protect ocean areas under multiple jurisdiction (Johnston and Enomoto, 1981; Remond-Gouilloud, 1981c). Finally, nations now have a duty, individually and collectively, to safeguard those ocean areas beyond national sovereignty (Fisheries Jurisdiction Case--United Kingdom v. Ireland, 1974; d'Amato and Hargrove, 1975).

This new doctrine of international law has been codified by the text of the proposed United Nations Convention on the Law of the Sea, in provisions dealing with pollution control, living marine resource management, and protection of the ocean environment. Those provisions are generally accepted by all nations, including those like the United States, which have not signed it because of concern with other provisions (Hoyle, 1983; Proclamation No. 5030, 1983; Malone, 1984).

In the proposed Law of the Sea Convention, primary responsibility for the control of pollution still remains with the coastal state for activities within their Exclusive Economic Zone (EEZ), with the flag states for their vessels wherever they may be, and with the port state for vessels docking on its shores (Remond-Gouilloud, 1981a; Schneider, 1982; Churchill and Lowe, 1983). However, with this responsibility are now placed obligations. Flag states are to adopt regulations for their vessels which "at least have the same effect as that of generally accepted international rules and standards" and must enforce them. Port states have the responsibility to enforce international standards for pollution control as to a vessel's activities in the high seas when the vessel responsible for such pollution enters their harbors (Arts. 211(2), 217-220 UNCLOS, 1982). Similarly, coastal states must adopt laws and regulations "to prevent, reduce and control pollution of the marine environment" from land-based sources, from activities occurring in their exclusive economic zone, including activities on their seabed, and from ocean dumping (Arts. 194, 207-211, 213-216, 220 UNCLOS, 1982). The Treaty expects controls to be implemented through domestic laws, bilateral and multilateral treaties, and other cooperative arrangements (Arts. 213 to 220 UNCLOS, 1982). Policies are to be harmonized at the regional level, and the goal is to establish appropriate global rules that take into account unique regional features (Art. 197 UNCLOS, 1982).

Similar rules are evolving as to management of living marine resources. As the result of expanded fisheries zones, bilateral and multilateral agreements have been ratified for fishing. Most of these are species specific and focus on protection of a threatened or endangered species or on limits to the capture of fish by nationals or a particular state. As with pollution standards, however, the international community, based on state practice and the new acceptance of the oceans as a commons, has been moving towards recognition of a customary rule of law governing fisheries conservation and management, which has been codified in the universally-accepted provisions of the proposed United Nations Convention on the Law of the Sea.

The Treaty confirms the new national jurisdiction over an adjacent 200-mile exclusive economic zone. Within this zone, coastal states have sovereignty over the exploitation of resources, and responsibility for the protection of the marine environment. The Treaty imposes a responsibility on the coastal state to conserve the living resources of the zone and to manage these resources. Coastal states must ensure that the living resources in the EEZ are not endangered by overexploitation, and that the population of harvested species are maintained or restored at maximum sustainable yields (Art. 61 UNCLOS, 1982). The provisions, therefore, imply that adequate national laws must be put into place to live up to these international responsibilities and that adequate scientific and statistical data must be kept to determine maximum yield and to warn of overexploitation (de Klemm, 1981).

A second feature of the Convention is to impose responsibilities on coastal states for transboundary resources. States are commanded to cooperate, through formal or informal agreements, to manage stocks which occur within the EEZs of two or more states. Similarly, the Treaty establishes international responsibility for management of species that occur within one or more EEZs and the high seas. Coastal states are to work with the home nations of fishermen to design rules for the fishing of stock within and without an EEZ. In addition, nation-states are to join together to provide a common set of rules, through international organizations, for highly migratory species, and for anadromous and catagromous species, that spend at least part of their lives in an EEZ (Arts. 63-64, 66-67 UNCLOS, 1982).

The establishment of such rules, through bilateral or multilateral agreements, could govern the management of 80 to 90 percent of the world's fishing (Alexander, 1983). In addition, the Convention also provides general standards for fishing management on the high seas. For such fishing, all states have the duty to exact such controls over their nationals as well as provide for the conservation of living resources in the areas of the high seas and to work with other states to develop management measures for such species. As with rules for EEZ management, high seas exploitation rules should include provisions to maintain or restore harvested species at levels that will assure maximum sustainable yields (Arts. 117 to 119 UNCLOS, 1982).

This overview of the new rules of customary international law indicates that the needed legal framework provides an opportunity for comprehensive management. It does not mandate it. The provisions of the Law of the Sea Treaty still focus basically on species specific management regimes and separate rules for control of marine pollution from those for conservation and management of living resources. This is the result of the nature of international law. Nation-states continue to see the need to protect the economic interests of their own citizens and to guard zealously their sovereign territorial independence and rights. The lack of comprehensive and enforceable controls are the result of these political realities. Controls will only occur as those states see the mutual benefit of establishing such rules and, thus, usually will occur only when a particular problem arises. As a result, the resolution of that particular problem is the focus of national or multinational action. The countries are not concerned with management of a total ecosystem (Pearson, 1982; Pardo, 1983; Springer, 1983). The next section of this chapter suggests options for the future based on that political reality.

Options for the Future

In an ideal world, international law could perhaps command total ecosystem management and have institutions to enforce those commands. However, such expectations of the law are unrealistic (McDougal, 1967). Effective total ecosystem management will continue to be affected, as with all other international laws and standards, by the political realities involved in global action. Although international law is evolving towards recognition of the oceans as a commons and consideration of the total and interrelated scope of oceans problems, it has not yet reached the point where anything more than moral support can be found for total ecosystem management.

The consensus of the international community on the non-deep seabed provisions of the proposed Law of the Sea Convention can be used to convert this moral support into new customary rules of international law. The Law of the Sea (LOS) treaty imposes responsibilities on states to protect the environment and manage fisheries. It can be urged that protection of all species and their habitats, and protection of the marine environment in general, can only be accomplished by total ecosystem management for large marine ecosystems. It is thus an implied next step in the evolving international law governing the oceans. The opportunity is present to use this new heightened environmental sensitivity to produce mechanisms for total ecosystem management. The key is flexibility and experimentation, with a few successes. Thus, while some efforts can be made to move all fisheries and pollution regimes to ones of comprehensive management, a concentration by the scientific community on a few areas of potential success are more likely to lead to international acceptance of total ecosystem management. The energy of those seeking comprehensive regimes should be focused at the two ends of the spectrum--large systems primarily within one or more EEZs and large systems with unique jurisdictional and scientific histories.

ECOSYSTEM MANAGEMENT WITHIN EEZs

It is estimated that 38% of the ocean and at least 75% to 80% of the potential world catch exists within the collective EEZs of all nations (Alexander, 1983; Churchill and Lowe, 1983). Scientists should identify ecosystems within one or more EEZs and work with government officials in such states to help design unilateral, or more often bilateral and multilateral, fishery management regimes that focus on total ecosystem management. The legal support for such comprehensive action is present. States have the authority to control all acts within their sovereign waters and to set such control on an ecosystem basis. Morover, the new rules for the Law of the Sea urge states to act, individually and collectively, to protect the oceans and their living resources. This can best be done through total ecosystem management (Cicin-Sain and Knecht, 1983).

The political support for such action is growing but must be developed. Nation-states still need to be convinced that total ecosystem management can work-- that is, that it can satisfy local resource needs and national and international requirements for protection of the marine environment and its resouces. Scientists and fisheries managers must exercise their political

254

power and convince their local decision-makers that
ecosystem management is first, a sound scientific
concept, and second, appropriate for selected large
marine ecosystems. Successful application in a few
cases, in nation-state fisheries plans, or in a few
bilateral or miltilateral fishing agreements, can lead
to further applications (Gordon, 1981).
    Selected EEZ application is, therefore, a real
testing ground. It can only work if government
leaders, scientists, and fishery managers design and
apply comprehensive approaches in a few illustrative
cases and thus change the way we think and act about
the oceans--from a species specific to an ecosystem
approach, and from separate fisheries and pollution
regimes to a regional seas comprehensive perspective
(United States Department of State and Council on
Environmental Quality, 1981).

ANTARCTICA

    While more localized efforts are attempted, the
international community should also seize the opportunity to set legal precedent in the new living
resource regime for Antarctia. Because of its unique
history of cooperative nation-state action, and environmental protection, the management of Antarctica's
resources can be an international demonstration of the
possibility and potential of total ecosystem management
(Mitchell and Sandbrook, 1980). In fact, the effort to
use Antarctica as a model for future management regimes
has already begun. In the Convention for the Conservation of Antarctic Marine Living Resources, management
is based on a total ecosystem conservation standard,
rather than on harvested target species. In addition,
the Convention requires signatory states to conduct
their affairs and watch over the affairs of others, so
as to minimize the risks to the Antarctic marine
ecosystem (Preamble and Arts. II(3), V, XXI, XXII,
Antarctic Living Marine Resources Convention, 1980).
As with any treaty, there are political problems and
conflicts of national interest with international cooperation that can hinder the success of the Antarctic
regime (Bilder, 1982; Zumberge, 1982; Boczek, 1983).
Still, if the effort is successful, it could convince
political leaders that total ecosystem management can
work for large marine ecosystems and that states can
satisfy and integrate domestic economic concerns into a
comprehensive multinational approach. As such, it
could be another step in the creation of an international customary law norm requiring total ecosystem
management (Barnes, 1982). It is therefore essential
that environmental interest groups, scientists, and

fisheries managers desiring international legal adoption of a total ecosystem approach support the Antarctic regime and work together to make it successful.

CONCLUSION

With the new enlightened international political sensitivity to total ecosystem management, there is the potential for establishing new international standards providing for application of the comprehensive approach to govern man's activities in the oceans. Such new international law will not result from the establishment of legislative-type world-wide rules. The nature of international law precludes such international mandates. Nation-states will continue to protect their own sovereign interests and international law supports this concept of national power. States will oppose new legal norms that diminish this power and will therefore agree to cooperative efforts only when they are convinced that such efforts benefit them.

The nature of international law, however, does provide the opportunity for the emergence of new customary rules, based on state and multistate practice as well as international acceptance. The more often individual states apply total ecosystem approaches both to their own management regimes and to those regimes they jointly control with other states, the more total ecosystem management will become the preferred, and therefore customary, rule of law. This trend can be confirmed by a few international efforts to design guidelines and to apply those guidelines in a total ecosystem approach in particular unique situations, such as in Antarctic marine waters.

Any constraints on total ecosystem management are not legal, but rather political. The legal options for such an approach are virtually limitless. The opportunities for application and acceptance of such comprehensive response to man's actions in the oceans must be carefully nurtured so as to allow a new rule of law favoring total ecosystem management of large marine ecosystems.

REFERENCES

Articles and Books

Alexander, L. M. 1967. Offshore claim of the world. In The law of the sea. pp. 71-84. Ed. by L. M. Alexander. The Ohio State University Press, Columbus, Ohio. 321 pp.
Alexander, L. M. 1983. The ocean enclosure movement: Inventory and prospect. San Diego L. Rev. 20:561, 580.

256

Barnes, J. 1982. The emerging Convention on the Conservation of Antarctic Marine Living Resources:
An attempt to meet the new reality of resource
exploitation in the Southern Ocean. In The
new nationalism and the use of common spaces.
pp. 239-286. Ed. by J. Charney. Allanheld, Osmun
& Co. Publishers, Inc., N.J. 343 pp.
Bender, K. 1981. Marine environmental protection in
the Scandinavian countries. In Comparative
marine policy. pp. 179-185. Center for Ocean
Management Studies. J. F. Bergin Publishers,
N.Y. 260 pp.
Bilder, R. B. 1982. The present legal and political
situation in Antarctica. In The new nationalism
and the use of common spaces. pp. 167-205. Ed.
by J. Charney. Allanheld, Osmun, & Co.
Publishers, Inc., N.J. 343 pp.
Boczek, B. 1983. The protection of the Antarctic
ecosystem: A study in international environmental
law. Ocean Dev. Int. L. 13:347.
Boxer, B. 1983. The Mediterranean Sea: Preparing and
implementing a regional act plan. In Environ-
mental protection. pp. 267-309. Ed. by D. Kay
and H. Jacobson. Allanheld, Osmun & Co. Publish-
ers, Inc., N.J. 340 pp.
Brierley, J. 1963. The law of nations. Clarendon
Press, Oxford. 442 pp.
Burke, W. T. 1967. Law and the new technologies.
In The law of the sea. pp. 204-227. Ed. by L.
Alexander. The Ohio State University Press,
Columbus, Ohio. 321 pp.
Burke, W. T. 1983. Scientific research. In The
United States without the Law of the Sea Treaty:
Opportunity and costs. pp. 113-115. Center for
Ocean Management Studies. Times Press Educational
Publishing, Providence, R. I. 271 pp.
Cicin-Sain, B. and Knecht, R. 1983. Implementing the
U.S. Exclusive Economic Zone: An opportunity for
improving ocean governance. In Assessing ocean
governance: Report of the First Meeting of the
Ocean Policy Roundtable. p. 15.
Chapman, W. M. 1967. Fishery resources in offshore
waters. In The law of the sea. pp. 87-105. Ed.
by L. Alexander. The Ohio State University Press,
Columbus, Ohio. 321 pp.
Chapman, W. M. 1970. The theory and practice of
international fishery development--management.
San Diego L. Rev. 7:408.
Chasis, S. 1981. Marine protection in the United
States. In Comparative marine policy. pp. 187-
194. Center for Ocean Management Studies. J. F.
Bergin Publishers, N.Y. 260 pp.

Churchill, R. and Lowe, A. 1983. The law of the
    sea. Manchester University Press, New Hamp-
    shire. 321 pp.
Comment. 1977. An environmental assessment of
    emerging fishery doctrine. Columbia J. Environ.
    Law 4:143.
Comment. 1983. American ocean policy adrift: An
    Exclusive Economic Zone as an alternative to the
    Law of the Sea Treaty. U. Fla. L. Rev. 35:492.
Comment. 1983. Recent developments in the Law of the
    Sea. San Diego L. Rev. 20:679, 706-08.
d'Amato, A. and Hargrove, J. 1975. An overview of the
    problem. In: Who protects the ocean? pp. 1-
    35. Ed. by J. Hargrove. West Publishing Co., St.
    Paul. 250 pp.
de Klemm, C. 1981. Living resources of the ocean.
    In The environmental law of the sea. pp. 71-
    192. Ed. by D. Johnston. International Union for
    Conservation of Nature and Natural Resources.
    Gland, Switzerland. 419 pp.
Farnell, J. 1981. EEC fisheries management policy.
    In Comparative marine policy. pp. 137-144.
    Center for Ocean Management Studies. J. F. Bergin
    Publishers, N.Y. 260 pp.
Friedheim, R. 1975. Ocean ecology and the world po-
    litical system. In Who protects the ocean? pp.
    151-190. Ed. by J. Hargrove. West Publishing
    Col, St. Paul. 250 pp.
Goldberg, E. et al. 1975. The role of enforcement
    analysis in international fishing regulation. pp.
    183-211. In The future of international fish-
    eries management. Ed. by G. Knight. West
    Publishing Co., St. Paul. 253 pp.
Goldie, L. F. E. 1975. International maritime envi-
    ronmental law today--an appraisal. In Who
    protects the ocean? pp. 63-121. Ed. by J.
    Hargrove. West Publishing Co., St Paul. 250 pp.
Gordon, W. 1981. Management of living marine re-
    sources: Challenge of the future. In Com-
    parative marine policy. pp. 195-167. Center for
    Ocean Management Studies. J. F. Bergin Pub-
    lishers, N.Y. 260 pp.
Greenberg, E. and Shapiro, M. E. 1982. Federalism in
    the Fishery Conservation Zone: A new rule for the
    states in an era of federal regulatory reform.
    So. Calif. L. Rev. 55:641.
Hargrove, J. 1975. Environment and the third con-
    ference on law of the sea. In Who protects the
    ocean? pp. 191-233. Ed. by J. Hargrove. West
    Publishing Co., St. Paul. 250 pp.
Hollick, A. 1981. U. S. foreign policy and the law of
    the sea. Princeton Univ. Press, N.J. 496 pp.

Hoyle, B. 1983. Comments. In The United States without the Law of the Sea Treaty: Opportunity and costs. pp. 99-101. Center for Ocean Management Studies. Times Press Educational Publishing, Providence, R.I. 271 pp.

Jacobson, J. and Davis, K. 1983. Federal fisheries management: Guide book to the Fishery Conservation and Management Act. The Oregon State University Sea Grant College Program, Corvallis, Oregon. 139 pp.

Johnston, D. and Enomoto, L. 1981. Regional approaches to the protection of the marine environment. In The environmental law of the sea. pp. 285-385. Ed. by D. Johnston. International Union for Conservation of Nature and Natural Resources. Gland, Switzerland. 419 pp.

Keckes, S. 1981. Regional seas: An emerging marine policy approach. In Comparative marine policy. pp. 17-20. Center for Ocean Management Studies. J. F. Bergin Publishers, N.Y. 260 pp.

Kildow, J. 1982. Political and economic dimensions of the land based sources in Marine Pollution. In: The new nationalism and the use of common spaces. pp. 68-89. Ed. by J. Charney. Allanheld, Osmun & Co. Publishers, Inc., N.J. 343 pp.

Knight, G. 1975. International fisheries management: A background paper. In The future of international fisheries management. pp.1-49. Ed. by G. Knight. West Publishing Co., St. Paul. 253 pp.

Larson, K. 1983. Case study: United States tuna policy. In The United States without the Law of the Sea Treaty. Center for Ocean Management Studies. pp. 213-215. Times Press Educational Publishing, Providence, R.I. 271 pp.

Lutz, R. E. 1976. The laws of environmental management: A comparative study. Am. J. Comp. L. 24:447.

Magnuson, W. G. 1977. The Fisheries Conservation and Management Act of 1976: First step toward improving management of marine fisheries. Wash. L. Rev. 52:427.

Malone, J. 1984. Who needs the sea treaty? Foreign Affairs 45:44.

McDougal, M. S. 1967. International law and the law of the sea. In the law of the sea. pp. 3-25. Ed. by L. M. Alexander. The Ohio State University Press, Ohio. 321 pp.

McManus, R. J. 1982. Legal aspects of land-based sources of pollution. In The new nationalism and the use of common spaces. pp. 90-111. Ed. by J. Charney. Allanheld, Osmun & Co. Publishers, Inc., N.J. 343 pp.

Mitchell, B., and Sandbrook, R. 1980. The management
of the Southern Ocean. Int. Inst. Environ. Dev.,
London. 162 pp.

National Advisory Committee on Oceans and Atmosphere.
1982. Fisheries for the future. U.S. Gov. Print.
Office, Washington, D.C. 61 pp.

Pardo, A. 1983. The Convention on the Law of the Sea:
A preliminary appraisal. San Diego L. Rev. 20:489.

Pearson, C. 1982. Environment and international economic policy. In environment and trade. pp. 46-
53. Ed. by S. Rubin and T. Graham. Allanheld
Osmun & Co. Publishers. N.J. 209 pp.

Remond-Gouilloud, M. 1981a. Land-based pollution.
In The environmental law of the sea. pp. 230-
245. Ed. by D. Johnston. International Union for
Conservation of nature and natural resources.
Gland, Switzerland. 419 pp.

Remond-Gouilloud, M. 1981b. Pollution from seabed
activities. In The environmental law of the
sea. pp. 245-258. Ed. by D. Johnston.
International Union for Conservation of Nature and
Natural Resources. Gland, Switzerland. 119 pp.

Remond-Gouilloud, M. 1981c. Prevention and control of
marine pollution. In The environmental law of
the sea. pp. 193-202. Ed. by D. Johnston.
International Union for Conservation of Nature and
Natural Resources. Gland, Switzerland. 419 pp.

Ross, D. 1982. Introduction to oceanography.
Prentiss-Hall, Inc. N.J. 544 pp.

Schaefer, M. 1970. Some recent developments concerning fishing and the conservation of the living
resources of the high seas. San Diego L. Rev.
7:371.

Schneider, J. 1981. Pollution by vessels. In The
environmental law of the sea. pp. 203-216. Ed.
by D. Johnston. International Union for Conservation of Nature and Natural Resources. Gland,
Switzerland.

Schneider, J. 1982. Prevention of pollution from
vessels or don't give up the ship. In The new
nationalism and the use of common spaces. pp. 7-
28. Ed. by J. Charney. Allanheld, Osmun & Co.
Publishers, Inc. 343 pp.

Sohn, L. 1973. The Stockholm declaration on the human
environment. Harv. Int. L. J. 14:423.

Springer, A. 1983. The international law of pollution. Quorum Books. Conn. 218 pp.

Travalio, G., and Clement, R. 1979. International
protection of marine mammals. Columbia J.
Environ. L. 5:199.

United States Department of State and Council on
Environmental Quality. 1981. Global future:

Time to act. U.S. Gov. Printing Office Wash., D.C. 209 pp.

Waldichuck, M. 1982. An international perspective on global marine pollution. In Impact of marine pollution on society. pp. 37-77. Center for Ocean Management Studies. Praeger Publishers, N.Y. 313 pp.

Wallace, S. A., and Ratcliffe, T. L. 1983. Water pollution laws: Can they be cleaned up? Tulane L. Rev. 57:1344.

Whipple, W. 1982. Land-Based source of marine pollution and national controls. In The New Nationalism and the Use of Common Spaces. pp. 29-67. Ed. by J. Charney. Allanheld, Osmun & Co. Publishers, N.J. 343 pp.

Wilkinson, C., and Conner, D. 1983. The law of the Pacific salmon fishery: conservation and allocation of a transboundary common property Resource. Kansas L. Rev. 32:17.

Zumberge, J. H. 1982. Potential mineral resource availability and possible environmental problems in Antarctica. In The new nationalism and the use of common spaces. pp. 115-154. Ed. by J. Charney. Allanheld, Osmun & Co. Publishers, N.J. 343 pp.

Cases

Fisheries Jurisdiction Case - United Kingdom v. Ireland [1974] I.C.J. Rep. 33.

Lauritzen v. Larson, 345 U.S. 571 (1953).

Saudi Arabia v. Arabian American Oil Co. 24 Int. L. Rep. 117, 211-218 (1958).

United States v. Flores 389 U.S. 137 (1933).

Whitney v. Robertson, 124 U.S. 190 (1888).

Wildenhus's Case, 120 U.S. 1 (1887).

Other Materials

Convention on the Conservation of Antarctic Marine Living Resources, May 7, 1980. T.I.A.S. No. 8826, reprinted in 19 Int. Legal Materials 841 (1980), (Antarctic Living Marine Resources Convention).

Convention on the Continental Shelf, April 29, 1958, T.I.A.S. No. 5578.

Convention on the High Seas, April 29, 1958, T.I.A.S. No. 5200.

Convention on the Prevention of Marine Pollution by Dumping of Wastes and Other Matter (The London Dumping Convention), Dec. 29, 1972, T.I.A.S. No. 8165.

International Convention on Civil Liability for Oil
    Pollution Damage (The Liability Convention), 9
    Int. Legal Materials 45 (1970).
International Convention for the Prevention of
    Pollution of the Sea by Oil, 327 UNTS 3 (1971).
International Convention for the Prevention of
    Pollution from Ships (Marpol), 12 Int. Legal
    Materials 1319 (1973).
Magnuson Fisheries Conservation and Management Act of
    1976, 90 Stat. 331, 16 U.S.C.A. §§1801 et seq.
    (West Supp. 1983).
Ports and Waterways Safety Act 86 Stat. 424, 33 U.S.C.
    §§1221 et seq. (1972).
Proclamation No. 2667, 10 Fed. Reg. 12303 (1945),
    codified at 3 C.F.R. 67 (1943-48 Compilation).
Proclamation No. 2668, 10 Fed. Reg. 12304 (1945),
    codified at 3 C.F.R. 68 (1943-48 Compilation).
Proclamation No. 5030, 48 Fed. Reg. 10605 (1983),
    codified at 3 C.F.R. 5030, reprinted in 22 Int.
    Legal Materials 465 (1983).
Statement by the President on United States Ocean
    Policy, accompanying his Proclamation establishing
    an Exclusive Economic Zone, 19 Wkly. Comp. Pres.
    Doc. at 383 (March 14, 1983), reprinted in 22 Int.
    Legal Materials 464.
Statute of the International Court of Justice (1945).
United Nations Convention on the Law of the Sea
    (UNCLOS), U.N. Doc. A/Conf. .62/121, reprinted in
    21 Int. Legal Materials 1245 (1982).

13. Can Large Marine Ecosystems Be Managed for Optimum Yield?[1]

ABSTRACT

The question--"Can LMEs be managed for optimum yield?"--suggests that demonstration of feasibility and cost-effectiveness of management may not be sufficient to get managers to manage the resources effectively. The pros and the cons of the question are examined, with the conclusion that the chief impediment to management is the political one of making decisions on the distribution of wealth. If this impediment can be overcome, then the path towards improved management is clear.

INTRODUCTION

I have been assigned the task of answering the question "Can large marine ecosystems be managed for optimum yield?" This question derives from the apprehension that demonstrations of feasibility and cost-effectiveness may not be sufficient to get resource managers to manage the resources.

I share this apprehension. There is a significant gap between theory and practice--between scientists and managers--in the field of fisheries management. This is due not only to the great difficulties of understanding marine ecosystems (and of developing theoretical models) but also, and perhaps more importantly, to the peculiar characteristics of common property which cause special problems for the practice of management. In seeking to close this gap, I suggest that the task is only in small part that of demonstrating technical and scientific feasibility and cost-effectiveness. The more important part is that of creating the means and institutions that will permit the managers to manage the resources and to use the inputs of science. These means and institutions can be found essentially in the establishment of satisfactory rights over the use of

the resources. The problem is that the establishment
of such rights requires decisions on the distribution
of   wealth--decisions   that   are   basically   and
fundamentally political.
     This chapter begins with some definitions of the
elements in the question--"Can LMEs be managed for
optimal yield?"   It then briefly examines a negative
response to the question, as a means for identifying
some of the major impediments to the practice of management. From this, it turns to a positive response.
Although the conclusion is that the need for wealth
distribution decisions forms an immense barrier to
effective management of LMEs, it is suggested that once
those decisions are made the path towards better management is fairly clear.

DEFINITIONS

     With regard to definitions, it is assumed here
that a "large marine ecosystem" is one that falls
within the EEZs of two or more countries (and perhaps
beyond their EEZs as well). On this basis, I interpret
the management of an LME as being the management of
shared stocks.
     I have some problem, however, with the term "opti-
mum yield."   This term presumes that a yield--or a
level of catch--can be optimal, in itself. The infer-
ence that can be drawn from much of the literature
referring to the concept of optimum yield is that social and economic factors need to be taken into account
only in determining the yield and that once the yield
is determined, then it is a simple matter of adjusting
fishing effort so as to produce that yield.
     This approach, however, by focusing on the pro-
duct, rather than the process, is the source of considerable mischief.   With regard to man's interests--
the social and economics aspects--it is not the level
of catch that is important but the difference between
the benefits produced and the costs and difficulties
incurred in producing those benefits.
     The proper function of management is that of de-
termining the nature of the benefits and costs and in
allocating the amount and kinds of capital and labor so
as to maximize net benefits. Because of the common
property characteristic of fisheries, the benefits and
costs are particularly difficult to define. In addition to monetary values, they may include job security,
national pride, supplies of protein, enforcement
requirements, transaction costs, etc., all of which may
be more significantly affected by the choice of management measures than by the choice of level of yield.
Management for optimum use (as against optimum yield)
requires, therefore, particular attention to the tech-

niques for allocating the factor inputs of capital, labor and natural resources.

MANAGEMENT OF LMEs IS NOT POSSIBLE

In reviewing the record of management of LMEs, it is difficult to find much evidence of optimum use. In the North Sea and the Baltic, in spite of the decades of scientific inputs through ICES, stocks continue to be overfished in both biologic and economic terms. ICES recommended a limit to the catch of Baltic cod of 197,000 tons in 1981. But the International Baltic Sea Fishery Commission could only reach an agreement on 272,000 tons, and actual landings amounted to 380,000 tons. For the North Sea, after years of negotiations, a Common Fisheries Policy was finally adopted by the European Community in January 1983. To facilitate the reduction in fishing effort, the Community is providing grants of 76 million ECU for the period 1983-1986. But, simultaneously, under another regulation, the Community is providing more than twice that amount (156 million ECU) for activities including the construction of new vessels and modernization of old.

Similarly, the use of the resources appears to be far from the optimum in almost all of the LMEs discussed in this volume (e.g., North Sea, northeast U.S. coast, the Antarctic, California Current, El Niño region off Peru, Gulf of Alaska, East Bering Sea, and the Baltic).

Generally, the fundamental problem lies in the difficulties of the sharing states in reaching agreements on the distribution of wealth. This is not surprising in view of the fact that in most of these LMEs wealth is defined more in terms of jobs than in terms of economic revenues. No national negotiator will last long in his office by sacrificing the employment opportunities of his constituents.

To sum up the negative side of the question, optimum use requires measures that permit effective allocations of capital, labor and natural resources. This, in turn, requires decisions on the distribution of wealth among the sharing countries. These decisions, since they generally affect jobs, cannot easily be made. Therefore, LMEs cannot "easily" be managed for optimum use.

MANAGEMENT OF LMEs IS POSSIBLE

There is some evidence to support a positive response to the question. Although not an LME as such, the fur seal fishery of the North Pacific was a shared stock that was optimally used. The decision was made, in 1909, to prohibit pelagic sealing and allow harvest

266

only on the breeding islands, where the costs of harvesting were lowest. This decision was possible only
because of a decision to share the wealth, by which
those countries which gave up pelagic sealing (Japan
and Canada) received a share of products from those
that did the harvesting (the U.S. and USSR). Although
the problems of the relationship between seals and
salmon were never resolved, the seals at least were
optimally used.

A more significant and exciting demonstration of
effective management is now taking place in the western
central Pacific. Here, the South Pacific Forum
Fisheries Agency, including 14 member and 2 observer
states, is making great strides towards regional
cooperation in the management of shared stocks for
tunas. The management measures being adopted by the
states are producing greater revenues from foreign
fishermen; permitting cost-effective monitoring,
control, and surveillance; and creating stable and
uniform conditions that benefit the foreign fishermen. It is noteworthy that these steps are being taken
in the absence of important scientific information such
as the effect of taking young yellowfin tuna by purse
seines on the yield of larger yellowfin taken by long-
lines; or the effect of fishing for juvenile skipjack
in the Philippines and Indonesia on the yield of mature
skipjack further east.

The success achieved thus far is due in large part
to the ease of making distribution decisions. The
skipjack resource is not overfished and little fishing
is done by the coastal states. It is not, therefore, a
question of distributing jobs but a question of distributing revenues from foreigners and sharing in the
costs of management.

To sum up the positive side, LMEs can be managed
for optimum use where wealth distribution decisions can
easily be made.

A HYPOTHETICAL PATH

The conclusion is that wealth distribution decisions are fundamental. No matter how clearly it is
demonstrated that management is technically and scientifically feasible and cost-effective, optimum use will
not occur in the absence of decisions on "who gets
what." These are political decisions among nations,
made in the context of a whole array of political
issues including many other matters than fisheries.

However, it can be hypothesized that once such
decisions are made, the path towards better management
is fairly clear, though perhaps quite long. International distribution decisions set the basis for distribution decisions at the national level. An indi-

vidual country whose fishermen acquire a right to take
a certain number of tons of fish is then in a position
to allocate shares of the quota among its fishermen or
to limit fishing effort. Once this is achieved, the
national fishermen acquire a form of property right in
the resource. This, in turn, can induce a proprietary
interest in the resource, facilitating compliance with
the regime. With the proper encouragement, the fishermen will find it within their interests to rationalize their efforts. This, in turn, sets the basis for
the fishermen assuming greater responsibilities for
management. Eventually along this path, the function
of management will be largely transferred from international bodies through national administrators and
into the hands of the fishermen where the decisions on
the allocation of capital and labor can best be made.

Once the fishermen acquire control over the means
of production, they will have the incentive to invest
in and heed scientific advice.

This hypothetical path is loaded with pitfalls,
diversions, and booby-traps. But there is sufficient
evidence to suggest that fishermen, given some degree
of property rights and left to themselves, will tend to
work towards optimum use of the resources. This has
been demonstrated in many studies of traditional
management systems of small marine ecosystems (e.g.,
lagoons and coral reefs).

SUMMARY

This discussion has largely begged the question
that I was asked. Although my answer is in the affir-
mative, it hinges on the ability of the participants in
an LME to make the necessary distributions decisions.
I have offered no suggestions as to how such decisions
can be made. Research on different approaches to
distribution decisions will be helpful. But,
ultimately, the decisions will be made when the
political costs of doing nothing are greater than the
political costs of dividing up the pie.

NOTES

1. The views expressed in this paper are those of
the author and do not necessarily reflect the views of
FAO.

14. Cost Benefit of Measuring Resource Variability in Large Marine Ecosystems

INTRODUCTION

Caveat

The difficulties encountered by blind men in describing an elephant are suggestive of those faced by an economist confronted by a large marine ecosystem (LME). In economics one can describe and analyze the entire market economy with a general equilibrium model, or deal with the problems of economic growth and business cycles with a macroeconomic model. At a less aggregated level for industry analysis there is partial equilibrium analysis with the boundaries of the industry definable in terms of elasticities. However, without an economic definition of the entity to be examined (an LME), not only is no general theory of the economics of LMEs possible but, even on a partial basis, one is limited to analysis of one specific relationship--here certain definable aspects of the interaction between economics and biology. Furthermore, examination of the interaction between biology and economics is limited in this paper to one instance in one geographic area, the fisheries of Georges Bank. On a more hopeful note, the problems revealed by examination of this one situation appear to be generally useful in considering the complex problem of the management of fisheries in the context of LMEs. Therefore, the analyses may apply to other multispecies fisheries, even though they involve other underlying differences in and linkages among the physical, chemical, biological, and economic parameters of LMEs.

A final heroic leap in the structure of this chapter is the use of the term "economic." As used here, economic stands not only for purely economic phenomena, such as the return on investment, but for the full range of economic, administrative, sociological, and psychological forces involved in the management of multispecies fisheries.

The Early Model

The first serious discussion of fishery management problems was over the "trawler question" in Britain in the late 19th century. This is worth noting as it was a discussion about aggregate yield, the question of diminishing returns in LMEs. Subsequent extensive and intensive growth in fisheries precluded further systematic analysis of this problem until the last half of this century.

The concept of diminishing returns--variable proportions--is a static construct describing the impact on output (yield) as varying amounts of inputs (labor) are applied to a fixed factor (land). In the fisheries, this would combine a year class (the only stock available for the fishery) with the variable input, fishing effort, to produce catch--catch that would in a year's time consume the fixed factor, the stock of fish. In this instance, if the inputs were not reduced concomitantly with the fixed factor, declining yields per unit of effort would be observed as a result of the combination of variable proportions and the reduction of the fixed factor.

However, the behavior of the economic agents is controlled by the combination of price and cost of output, so that their response to declining yields is a priori indeterminant. In a multispecies fishery under dynamic conditions with continuous year classes of different sizes entering the fishery, inventory surviving from previous year classes and different species with different economic value, the condition of particular year classes and the status of variable proportion and stock size are concealed. This circumstance is further complicated by the possibility of geographic expansion of the fishery, a possibility that has been central in fishery development in this century. However, with the possibility of geographic extension limited as a result of institutional change (extended economic zones) and high rates of exploitation of valuable stocks, the management of fisheries is having to face the control theory problem buried in the management of multispecies fisheries in LMEs.

When the question of fisheries management emerged in this century it focused on the physical conservation of specific high-valued species--salmon, halibut, etc. The development by biologists in the 1940s and 1950s of models of the "population dynamics" of individual fish stocks was an "open sesame" to economists. Economists took from the biologist the quantified yield functions as economic production functions and went on to develop simple models that optimized the yield from the resource. The definition by the biologists of the maximum sustainable yield (MSY) for a specific stock

led directly to the definition of the net economic
yield (NEY), a definition which incorporated price and
cost in the model. These models were, it appeared,
sufficient for the purpose of the rational economic
management of fisheries. They were deemed sufficient
since they both optimized the use of the economic
inputs and met the conservation objective of the
biologist (Pontecorvo, 1966).

In following this path, the economists walked into
a trap. The assumption made by the economists was that
the biologist's yield curve was a simple production
relationship of the type they were familiar with. This
involved an implicit assumption that the function was
stable with a small variance. If this had been so, a
basic problem, uncertainty about the yield, would have
been eliminated and the management of fisheries would
be simplified. Unfortunately it took several decades
to really understand that a Schaeffer, Beverton and
Holt, etc., type yield function was both unstable and
of equal economic importance, and that regardless of
the nature of the stability properties of the model,
the variance associated with the function was large.
We spent many years trying to manage fisheries on the
assumption that the MSY was definable and stable, and
therefore that it and the NEY were the appropriate
management objectives. Hopefully, our current level of
understanding is such that we all, biologists and economists, recognize that no simple optimizing solution to
the problems of fisheries management is at hand.

> The aim to achieve certainty and precision
> through quantification of the biological sys-
> tem seems rather to have revealed a complex
> stochastic network of relations in which the
> uncertainty and changing values of the
> parameters are reminiscent of the situations
> studied by economists [Dickie, 1979; for an
> earlier discussion of the stability of MSY
> see Pontecorvo, 1971].

RISK, UNCERTAINTY, AND OPTIMIZATION

Equilibrium

Equilibrium in a simple bioeconomic model of a
high-valued single species fishery under common property conditions (free and easy entry) involved the
catch equal to the MSY, with total revenue to the fleet
equal to the cost of production, and with the fleet
having some excess capacity in capital and labor whose
opportunity earnings are included in the aggregate cost
of production. If effort was restricted to the level
of the NEY, the equilibirum would have no excess capital and labor.

In this model the fishermen-vessel owners face three distinct risks:[1]

(1) The risk of change in the availability of fish and therefore the catch; i.e., there is a high level of variance associated with MSY, or if the yield function is unstable, or both. In these circumstances the fishermen may reap a bonanza or go bankrupt if they have high fixed costs, e.g., heavy finance charges, etc. Since the probability of a high variance in a fish stock is often increased by the intensity of fishing effort, the activities of fishermen may contribute to the risks they run.

(2) The economic risk of a change in price for the end product, or a change in the cost of production, or both. This risk may stem from exogenous causes (the normal behavior of the business cycle) or short-run endogeneous conditions (excess supply or demand).

(3) The third risk is the risk associated with regulation. Any change in the regulations a fisherman must conform to imposes a cost and, quite possibly, a loss in revenue.

There is an additional hidden difficulty facing the fisherman--the variability in an LME; i.e., the risk associated with the variability in the LME which reveals itself to the fisherman as the first risk, a change in the abundance of a particular stock which is an element in the LME. The duality of this problem will be discussed below, but as a first approximation one may consider it as a broader risk which may have an impact on the revenue and behavior of fishermen.

Finally, if we compare the fisherman with the farmer we note an important difference: for the marginal farmer, as for the average fisherman, costs equal revenue, but all intramarginal farmers will earn and keep intramarginal rents. Some exceptionally skilled fishermen will earn these rents, but most, due to the common property condition, will be at the margin. Therefore, economic agents in the fishery will tend to have fewer financial resources or borrowing capacity and accordingly, they are more vulnerable to all types of risk. The excess capacity in the fishery also results in financial weakness since easy entry tends to skim off earnings and, in the face of declining revenues, the redundant capital and labor remains in the fishery to claim a share of the falling income. This is an aspect of the familiar asymmetry between gains and losses in a market system. Profits, either real or expected, attract entry. Once entry takes place, for a time only variable costs need to be covered so that capital and labor will remain in a declining or unprofitable fishery in the expectation that ultimately adequate profits will be realized. (For a discussion of how this process has worked over an 80-yr period for the fisheries of Georges Bank see Pontecorvo, in press).

Disequilibrium

The sources of instability in an LME are complex. They involve the underlying physical and chemical influences within the structure of an LME as well as the biological and economic impacts. The implications of both the physical and chemical parameters and the linkages of those parameters to the biological aspects of an LME represent a set of issues that here we can only note.

The two remaining sources of instability in the LME, the biological and economic, combine to present, as Dickie (1979) suggested, a set of stochastic relationships that require the solution to a complex control theory problem in order to optimize the yield (Donaldson and Pontecorvo, 1980).

To optimize the economic yield from an LME two conditions must be met: The yield function for each stock in the biomass must be known and the NEY obtained from each stock. While theoretically this problem can be solved, as noted below, the solution is in all probability both uneconomic--the costs exceed the benefits--and administratively extremely difficult. These difficulties of cost and administration are reduced if the fishery is rationalized, i.e., it is efficient in production at a point in time. Further, over time it must continue as an efficient industry as productivity gains that result from technological changes reduce the inputs required per unit of output. However, as desirable as it is from the perspective of the welfare (size of GNP) of the nation, efficiency in production is not a necessary condition for simplification of the management of fisheries as discussed below. Simplification of the management process can be accomplished by a reduction in catch to a figure at or close to the lower level of the probable variance in each individual stock in an LME.

In this circumstance the trade-off is between a quantity of fish left in the ocean (this quantity to vary over time with the variation in the individual stocks and the variance in the LME), and lower costs of management of the fishery and a lower level of capital and labor inputs. A priori it is likely that a fishery managed with the catch at a biological sustainable minimum level of availability will be easier to rationalize.

Finally, there are price (potential increases) and therefore welfare implications involved in moving the physical level of catch to a mean value below the mean value of the potential MSY for each stock. Assuming

that the cost of management is reduced by the movement
to a lower level of catch and that the cost of
management is paid out of general tax revenues, then
the welfare impact of a change in the level of catch
can only be ascertained by reference to the tradeoff
between lower taxes and price increases in fish protein. This will require analysis of each specific LME
and the market for its output. This problem is further
complicated by the uncertain stock effects of the shift
from a higher to a lower level of fishing effort (Table
14.1; Beddington and Rettig, 1984).

Given predation by man and natural variation from
biological, chemical, and physical effects, there are
"n" possible biomass states. In one LME any biomass
state i may differ from state j for a variety of reasons. The change may come from underneath, i.e., the
physical and chemical aspects of the LME or from
biological or economic sources. As suggested in Table
14.1, the variation may be in the aggregate weight of
the biomass in the relative abundance of the stocks
which compose the biomass and in variation in the size
distribution of the elements in each stock.

Biomass variation has two economic implications.
The first is a problem in capital theory. If biomass
state i exists and state j is the preferred economic
condition, the economic cost of the transition from i
to j is the net present value of the yield foregone
(i.e., to increase the size of stock A and E the
industry must refrain from or reduce the catch of A and
E to allow the stocks to grow), in order to achieve the
alternative biomass, plus the cost of the regulatory
procedures required to change biomass states.

Biological change is not a simple problem in
production. It is not possible to measure the biological response mechanism as if it were a change in
the size of automobiles. This difficulty is rooted in
the biological recruitment process (Beddington and
Rettig, 1984; Beddington, this volume; Sissenwine, this
volume), so that while to the economist state j may be
preferred to i, there is no known way to measure either
the time required to move from i to j nor even if it is
possible to achieve state j. It is not impossible that
a reduction in effort planned to move the biomass from
i to j will result in a movement from i to some alternative state, n, rather than j.

Within each biomass state there will be
differences in the several dimensions of the individual
stocks. Changes in the biomass of each individual
stock, again, have a variety of causes. The most
obvious of these is from fishing effort, although in
some cases other predators may be important as well.
These predator/prey relationships may be complex; one
stock in the biomass may stand as part of the food

supply for another element in the biomass, and in turn
the juveniles of the predators may be the food supply
for other elements in the ecosystem, etc.

COSTS, BENEFITS, AND FISHERIES MANAGEMENT

The economics of variations in LMEs are centered
in the dynamics of aggregate biomass and stock biomass
change and in the risk associated with variation in
demand (Pontecorvo, in press). In the case of Georges
Bank, the supply situation is that a given biomass is
in place and that a yield of approximately 125,000 tons
of the existing mix of fin fish and shell fish represents the set of MSYs for the several stocks. The
range of variance is estimated as 100,000 to 150,000
tons. The second dimension of change, the movement in
individual stock biomasses, is more pronounced, e.g.,
the haddock stock on Georges Bank provided in a yield
of approximately 10,000 tons in the period 1904-1910,
115,000 tons in 1929, and 2000 to 3000 tons in the
early 1970s.

Therefore, even if today's biomass state is
preferred and if it is stable, it still presents to the
fisherman considerable risk in both year to year aggregate yield and in the yield of particular species.
Note that species are not perfect substitutes for each
other, i.e., the U.S. market, which is the basic market
for the output of the LME represented by Georges Bank,
pays a relatively high price for haddock and a very low
one for squid, so that variation by species presents a
substantial economic risk (Sissenwine, this volume).

To summarize the management problem, assume any
biomass state is the one preferred in economic terms,
i.e., if all stocks are fished to yield the net
economic yield--the present value of the net revenue
stream, will be maximized against all alternative
biomass states. There are, in these circumstances,
three costs to be considered:

(1) The high cost of the biological and economic
information which is required by the regulators of
fishing effort in order to achieve biomass stability;
one must continually monitor the condition of all the
elements in the biomass (including environmental
forces) to assure the yields (MSYs) are in equilibrium
and, in addition, current daily economic information on
prices and costs must be obtained as part of the optimization process (Beddington, this volume).

(2) The costs which derive from the normal variation in the catch; these costs are imposed upon the
fishermen both in terms of production costs and the
level of revenue. Furthermore, these costs are affected by the common property condition of the resource

276

TABLE 14.1.

LME (1) that contains Biomass State i consisting of five stocks A-E[a]

BIOMASS STATE i

   weight

STOCK A_i a few large fish many small fish	-- yield (MSY) a; (market price of a) - (minimum cost of production of a) = net economic yield, a_i
STOCK B_i a few large fish many small fish	-- yield (MSY) b; (market price of b) - (minimum cost of production of b) = net economic yield, b_i
STOCK C_i a few large fish many small fish	-- yield (MSY) c; (market price of c) - (minimum cost of production of c) = net economic yield, c_i
STOCK D_i a few large fish many small fish	-- yield (MSY) d; (market price of d) - (minimum cost of production of d) = net economic yield, d_i
STOCK E_i a few large fish many small fish	-- yield (MSY) e; (market price of e) - (minimum cost of production of e) = net economic yield, e_i

Aggregate weight of biomass i = Σ $(A_i + B_i + C_i + D_i + E_i)$

Net economic yield of biomass i = Σ $(a_i + b_i + c_i + d_i + e_i)$

---

[a] Considered as an asset, the value of any biomass state "n" is the net present value of the expected net economic yield.

TABLE 14.1. (Continued)

LME (1) where Biomass State i has been transformed to Biomass State j[a]

BIOMASS STATE j
   weight

STOCK Aj size of fish, intermediate	— yield (MSY) a; (market price of a) − (minimum cost of production of a) = net economic yield, a_j
STOCK Bj size of fish, intermediate	— yield (MSY) b; (market price of b) − (minimum cost of production of b) = net economic yield, b_j
STOCK Cj size of fish, intermediate	— yield (MSY) c; (market price of c) − (minimum cost of production of c) = net economic yield, c_j
STOCK Dj size of fish, intermediate	— yield (MSY) d; (market price of d) − (minimum cost of production of d) = net economic yield, d_j
STOCK Ej size of fish, intermediate	— yield (MSY) e; (market price of e) − (minimum cost of production of e) = net economic yield, e_j

Aggregate weight of biomass j = $\Sigma (A_j + B_j + C_j + D_j + E_j)$

where $AW_j > AW_i$

Net economic yield of biomass i = $\Sigma (a_j + b_j + c_j + d_j + e_j)$

[a] Considered as an asset, the value of any biomass state "n" is the net present value of the expected net economic yield.

which allows excess capital and labor in the fishery. Also, business cycles impact in the usual way upon demand and cost, and in an additional way on an inefficient industry with excess capital and labor; e.g., consider the impact of a downward shift in supply concomitant with a drop in price in an overcapitalized fishery.

(3)   Finally, there is the cost of adjustment, which is imposed on regulators and fishermen alike. This is both an economic cost and a human or psychological burden. Assume that all the information required to optimize the yield from the LME is available free. However, in order to optimize the yield the regulators must vary the mesh size of nets bimonthly and also the season must be opened and closed three times each year. The industry can, with difficulty, and if it is efficient (i.e., the common property problem has been rationalized) probably adjust to these changes.

However, if a complex regulatory system is based on models whose forecasting accuracy involves probabilities that are considerably less than one, then the continuous adjustments required of fishermen and regulators breaks down because of the problem of competitive withdrawal combined with industry doubts over the validity of any specific regulation that is imposed.

CONCLUSION

Variation, both natural and that imposed by human predation, in the biomass and in the stocks that constitute the elements of the biomass create economic risks for the exploiters of the resource. These economic risks are compounded by the impact of the common property problem on earnings, the rise and fall of the business cycle, exogenously created expectations, and inconsistent subsidy and tax programs.

On the management side there are two crucial issues: the cost of information and the necessity for stability in the regulator process. At the present time, and for the foreseeable future, it appears that the cost of both the quantity and quality of the information required to optimize the net economic yield from the biomass exceeds any possible benefits. Further, even if the required biological and economic information was freely available, the optimization process would impose, at considerable cost to both regulators and fishermen, constant changes in the pattern of exploitation.

In an open access fishery with redundant capital and labor, these changes would be intolerable; in one with limited entry they would make the regulator program difficult at best. In these circumstances the

appropriate policy for a multispecies fishery such as Georges Bank is a second best solution:
(1)   a general limitation on fishing effort to keep the catch at or close to the minimum level of the variance in the MSY for each stock in an attempt to stabilize the catch.
(2)   rationalization of the fishery to increase the financial strength of firms and to allow them to plan for the investment required to catch the reduced quantum of fish available to them.
(3) 1 and 2 represent a significant reduction in the information required and therefore a significant reduction in the cost of managing the fishery and also in the risks faced by the exploiters.

NOTES

1.   Risk as used here means uncertainty in the Knight (1921) definition.   That is the probability of events is not fully known, and therefore the risk is not insurable.

REFERENCES

Beddington, J. R.   This volume.   Shifts in resource populations in large marine ecosystems.
Beddington, J. R., and Rettig, R. B.   1984.   Approaches to the regulation of fishing effort.   FAO Fish. Tech. Pap. 243, pp. 1-4, 8-9.
Dickie, L. M.   1979.   Perspectives on Fisheries Biology and Implications for Management.   J. Fish. Res. Board Can. 36:838-844.
Donaldson, J., and Pontecorvo, G.   1980.   Economic rationalization of fisheries: The problem of conflicting national interests on Georges Bank. Ocean Development Int. Law J.   8(2):149-169.
Knight, F. H.   1921.   Risk, uncertainty, and profit. Chapter 2.   Houghton Mifflin Co., Boston, MA.
Pontecorvo, G.   1966.   Optimization and taxation, the case of an open access resources.   In The fishery, conservation, taxation and the public interest in extractive Resources, pp. 157-167.   Ed. by M. Gaffney.   Univ. Wisconsin Press, Madison, WI.
Pontecorvo, G.   1971.   On the Utility of Bio-economic Models for Fisheries Management.   U.S. Dep. Commer., NOAA Tech. Rep. NMFS CIRC-371, pp. 12-22.
Pontecorvo, G.   In press.   Supply, demand, and common property: The historical dynamics of the fisheries of Georges Bank; some preliminary observations.   Crutchfield-Festschrift, Univ. Washington Press, Seattle, WA.
Sissenwine, M. P.   This volume.   Perturbation of a predator-controlled continental shelf ecosystem.

R. Tucker Scully, William Y. Brown,
Bruce S. Manheim

15. The Convention for the Conservation of Antarctic Marine Living Resources: A Model for Large Marine Ecosystem Management

The Convention on the Conservation of Antarctic Marine Living Resources (CCAMLR), which entered into force in 1982, represents an effort to develop and apply an ecosystem-wide management approach to the waters surrounding Antarctica. CCAMLR is an innovative agreement. It applies to an area whose definition rests more on an ecological premise than upon political or economic considerations. It aims at ensuring that the harvesting activities are consistent with the health of not only target populations but also of the populations of dependent and related species and with the maintenance of ecological relationships.

Conclusion of CCAMLR, and in fact the nature of the Convention itself, rests in large part upon the cooperative international system established in Antarctica by the Antarctic Treaty of 1959. That Treaty provided the political mechanism upon which to develop CCAMLR as well as the catalyst for dealing with differences among the nations active in Antarctica over its legal and political status. Equally important, cooperative scientific research fostered by the Antarctic Treaty illuminated the need for and suggested the nature of a system to ensure conservation of living resources in Antarctic waters.

In the decades following the International Geophysical Year of the late 1950s, oceanographic and marine biological studies--building upon early pioneer research--led to the understanding that the waters surrounding Antarctica form a distinct marine region. This region defines itself both in oceanographic terms and in terms of a characteristic resident flora and fauna.

In physical terms, though the waters surrounding Antarctica are open to the major ocean basins to the north, there is a natural boundary which defines Antarctic waters (or Southern Ocean as these waters are often called). This boundary--the Antarctic Conver-

gence or Polar Front--is the complex transition zone
lying between 45° and 60° South Latitude within which
colder Antarctic waters sink beneath the warmer sub-
Antarctic waters to the north. Though there are important migratory populations--particularly of marine mammals and birds--which cross the Antarctic Convergence
seasonally, the marine area south of the Convergence
displays a characteristic assemblage of species. The
relationship among these species appears to be characterized by short, simple food chains, including dependence of most higher-order predators upon a single
species--Antarctic krill (Euphausia superba)--as a food
source.

The scientific research which outlined the large
marine ecosystem surrounding Antarctica also established the need for a mechanism to manage and ensure
proper conservation of the species found there. As the
nature of the ecological relationships in Antarctic
waters became better understood, so also grew the
conclusion that this ecosystem could be particularly
vulnerable to large-scale fishing. This recognition
was a prime impetus to the negotiation of CCAMLR for
seeking in that Convention application of an ecosystem-
wide approach to resource management. This approach is
reflected both in the definition of the area of the
Convention and in the conservation objective, set forth
in Article II of the Convention.

With respect to the area, it was recognized from
the outset that the requirements of ecosystem management would require coverage of an area larger than of
the the Antarctic Treaty (which applies to the area
south of 60° South Latitude). In deciding to initiate
formal negotiations, the Antarctic Treaty Consultative
Parties in 1977 agreed that the agreement should apply
not only to the area of the Antarctic Treaty but also
should "extend north of 60° South Latitude where that
is necessary for the effective conservation of species
of the Antarctic ecosystem . . . ."

In the negotiations themselves, the Parties
determined to define the area of the Antarctic marine
ecosystem by reference to the Antarctic Convergence
rather than on the basis of other geographic,
political, or economic factors. As a result, CCAMLR
applies to the area south of 60° South Latitude and to
the Antarctic marine living resources of the area
between that latitude and the Antarctic Convergence
which form a part of the Antarctic marine ecosystem.
Antarctic marine living resources are defined as the
populations of all species found south of the Antarctic
Convergence and the Antarctic marine ecosystem in turn
is defined as the complex of relationships of these
species with each other and with their physical
environment. Finally, the Antarctic Convergence is

deemed to be a line connecting specific geographic coordinates. It was recognized that a specific line cannot precisely depict the location of a boundary such as the Antarctic Convergence which is in fact a transition zone, shifting seasonally and in response to other environmental factors. At the same time the line was depicted in so far as possible to ensure that ·it included all areas south of the Convergence.

The concept of a large marine ecosystem requiring integrated management is the clear basis of defining the area of CCAMLR. Use of a natural oceanographic boundary represents use of the significant innovations incorporated into this agreement. The concept of seeking integrated management of a large marine ecosystem is also inherent in the conservation objective set forth in Article II.

Article II of the Convention sets forth 3 conservation principles that guide harvesting and associated activities south of the Antarctic Convergence. Although the Convention was negotiated in 1980 primarily in response to the likelihood of a greatly expanded unregulated commercial harvest of krill, these conservation principles are meant to be applied to all living organisms in the Convention Area. Perhaps the most important and most difficult task relating to the implementation of the Convention will, therefore, be interpretation of these conservation principles and their translation into criteria and methods for the formulation of conservation measures. Properly construed, however, these principles of conservation could serve as complementary guides in pursuit of the overall goal of preserving the Antarctic marine ecosystem while also allowing some degree of harvest.

The first conservation principle of Article II requires harvesting and associated activities to prevent a decrease in the size of any harvested population to levels below those which ensure the population's stable recruitment. For this purpose, the principle states, the size of a population should not be allowed to fall below a level close to that which ensures greatest net annual increment. This standard draws upon and is similar to the management regimes for the U.S. Marine Mammal Protection Act and the International Whaling Convention. The level of greatest net annual increment is the ultimate limit on harvest. If a population falls below this level, then harvest must be consistent with recovery to the level of greatest net annual increment or must cease. This ultimate limit on harvesting activities not only applies to krill, fish, and whales, which are principally the subject of harvest, but also to populations of other species that directly or indirectly interact with a harvested population.

Implementation of the first conservation principle may be somewhat problematic for it is widely believed that it is not possible to maximize the net annual increment of organisms at all levels of the food chain simultaneously. Moreover, in establishing acceptable harvest levels of krill, the parties to the Convention will be required to consider difficult questions about appropriate baseline levels for certain species. For example, the blue and humpback whales are today severely depleted when compared to their original population sizes. As a result of overharvest, the blue whale has declined from 200,000 to 10,000 individuals and the humpback whale from 100,000 to 3,000 individuals. Because of the relatively recent increased abundance of other krill predators such as minke whales, seals, penguins, and squid, however, these depleted whale species may not have access to the amount of krill that sustained their original populations. Hence, if viewed solely in terms of currently accessible krill, these endangered whale species could today be considered to be at or near the level of greatest net annual increment mandated by the Convention. Such a conclusion would be unacceptable, however, because it would reduce the requirements for restoration of such exploited populations in proportion to the degree of exploitation prior to establishment of a baseline.

The second conservation principle of Article II, to a limited extent, responds to the problems described above. It requires harvesting and associated activities to prevent changes or minimize the risk of changes in the marine ecosystem which are not potentially reversible over two to three decades. This principle recognizes that adaptive change may result from artificial selection as well as ecological change, and thus addresses the resilience of the Antarctic ecosystem to harvesting and associated activities. Hence, while the first principle sets limits on the degree to which populations may be altered by human exploitation, the second conservation principle sets a rate at which such changes must be reversible. For example, assuming the availability of krill remained at relatively constant levels, at least 30 years will be necessary for baleen whales at a level of greatest net annual increment to return to their initial population size after cessation of harvesting. If krill were overexploited, however, then that harvest could protract or even prevent recovery of the baleen whale population unless associated populations of competitors were manipulated. By setting limits on the degree to which populations may be altered by human exploitation, and setting a rate at which such alterations must be reversible, the Convention parties have endeavored to approximate the natural ecosystem while allowing for some degree of exploitation.

The third principle of Article II requires parties to the Convention to maintain the ecological relationships that exist between harvested, dependent, and related populations of Antarctic marine living resources, and to restore depleted populations to levels which ensure the greatest net annual increment. This provision, among other things, provides authorization for designating selected protected areas of sea, where harvest would be prohibited unless it would restore the ecosystem to such a structure and function as it was before harvesting occurred. This principle, therefore, can be implemented by extending to the Southern Ocean the notion widely accepted elsewhere that certain pristine areas should be set aside from exploitation of any kind. Establishment of such areas in the Southern Ocean, moreover, would provide a hedge against uncertainty and the risk of inadvertent exploitation in harvesting elsewhere.

Although the Antarctic ecosystem is broadly defined as that "complex of relationships of Antarctic marine living resources with each other and with their physical environment," it is clear that effective implementation of the conservation principles of Article II eventually will require subdivision of the Convention area into ecologically differentiated management areas. The Antarctic marine ecosystem can be subdivided into systems. For example, biological communities of open water can be distinguished from the community dependent upon pack ice areas. In addition, there is evidence that baleen whales may feed on the same krill concentrations from year to year rather than feeding freely throughout the Antarctic waters. Such "discrete" areas would be managed through an appropriate combination and application of conservation measures adopted pursuant to the principles of conservation set forth under Article II. Harvest efforts within these areas also could be regulated differentially to facilitate an analysis of the effects of different harvest regimes and ecological relationships between and among various species of Antarctic marine living organisms. The Convention parties also are called upon by the treaty to identify management indicator species and depleted species. These species should be sampled periodically for relevant life history statistics. The harvest of krill and other species must then be regulated in ecologically differentiated areas in a manner consistent with the maintenance and restoration of harvested species, indicator species, and depleted species to levels of greatest net annual increment.

To implement the conservation principles of Article II, the Convention established a Commission and

Scientific Committee with authority to impose a panoply of conservation measures and the flexibility to adopt and revise such measures expeditiously. Although these bodies resolved largely procedural matters during their first two meetings, this year the Convention parties adopted conservation measures to protect certain fish species from overharvest. At the same time, the Scientific Committee created two small working groups to undertake detailed scientific work over the next year. One group will analyze catch and effort data to determine age and length composition and recruitment trends of certain fish stocks, while the other will assess natural variations in the Antarctic ecosystem and investigate species related to and dependent on krill. The results of this work will form the basis for discussion and further implementation of Article II at the Convention's next annual meeting. Hence, it appears the Convention parties are on their way toward achieving the goal of the Convention--to protect the Antarctic through a comprehensive approach to ecosystem management. Although much remains to be done, it is clear, at least at this point, that the stage is set for a willing international community to establish a precedent of prudent use and conservation of one of the world's last unspoiled natural areas.

16. Very Large Ecosystems: From the Research Administrator's Point of View

ABSTRACT

   This chapter stresses two main points. The first
is that terrestrial and marine ecological paradigms are
different. The second point focuses on the fact that
there are different time scales of abiotic influence on
fish population processes, such as recruitment. More
study is required in this regard, particularly since
the abiotic factors that operate in the oceans exert
their influence on organisms more directly than on
land, and also provide considerable predictive
capability.

INTRODUCTION

   During my tenure as a research administrator in a
mission oriented institution, some general guidelines
have struck me as appropriate with respect to what one
does if the institution's mission is applied research,
as contrasted with what one might best do if the institution was charged with more basic studies in the marine sciences. These are:
   (1) to carry out the necessary, routine survey
type of activity that requires a continuing commitment,
and usually involves the use of a large research
vessel, and to encourage thoughtful and indepth
analysis of data collected beyond its immediate
practical use;
   (2) to maintain an up-to-date summary of all
knowledge in the relevant disciplines, including a
quality controlled data base for the use of one and
all;
   (3) to support basic research projects where
necessary to improve our capability to support the
mission, and within the fiscal and temporal flexibility
allowed, to encourage those who wish to do basic
research that is in the general area of the
institution's mission; but

(4) to avoid initiating basic research programs if they are being carried out in academia, or if they can be carried out efficiently and quickly enough by institutions and individuals that are charged with basic research activity.

As a research administrator you are, for all practical purposes, the final scientific arbiter of priorities, and, under the best of circumstances, not the originator of them. It is preferable to have the priorities develop "at the grass roots" as the time worn expression goes. That is often, although not always, the case. The research administrator must at least set the "tone" with objectives--very generally stated--and with a broad perspective. Such a tone may be set at staff meetings, or via institutional documents, scientific meetings, or publicatons (cf. Edwards, 1976).

How do you establish such objectives?

There are the obvious activities that provide background including endless program reviews, reading manuscripts, meetings of societies, symposia, and naturally discussions over coffee and beer. For myself, I have also found that reading diverse scientific literature, particularly literature which has nothing directly to do with fisheries, biological oceanography or other closely related material, to be most useful. There are some obvious benefits in the latter activity. It helps to keep results in your own area, so often reviewed and discussed, from becoming stale, and it serves to broaden one's perspective.

Personal experience certainly plays a major role. My own interest in large continental shelf ecosystems goes back a long way. My first position with the fisheries involved a study of the New England mixed fisheries--the "industrial" fishery of southern New England. It was an eclectic fishery with everything taken in the net being landed, either as fish for human consumption or reduction to meal. Since a relatively small-meshed net was used, the fishes taken presented an unusual opportunity to examine the seasonal mix of species, and their movements from one area to another.

It did not take very long, with such material at hand to develop an appreciation of the fact that while the ocean regimes were superficially very much like land ecosystems, the environmental clues that determined behavior and survival were markedly different from those to which terrestrial organisms responded. The animals operated in much the same manner, internally, as land animals, but the environmental rules and conditions were not the same.

For some obvious examples, fishes grow continuously and as they grow they occupy different niches in the ecosystem. You could say that they evolve from one species to another with each doubling of the size of their mouths. Microclimates and vegetation, almost the "sine qua non" of terrestrial ecosystems, are not characteristic of the sea. On land a bird or a squirrel can return to the same tree for food, nesting or shelter, but in the ocean an organism orients to the ambient dynamic physical aspects of the environment (Cf. Iles and Sinclair, 1982). Paradoxically, the environmental dynamics that fishes respond to can be more fixed geographically than the dynamics that control the limits of a particular vegetational association on land. An ocean ecosystem "hunts" stochastically for an equilibrium in terms of the composition of a region, but in not quite the same manner as on land, where for example, cycles and specific interactions between species seem to be more in evidence. That is not so surprising since fishes are extreme "r" strategists compared with land animals. Spawning stock size seemingly has relatively little to do with subsequent recruitment. After all, the adults are in a very different niche.

Referring back to the comment that the system hunts for an equilibrium randomly, it also is obvious that atmospheric phenomena play a subliminal, difficult to understand, but unmistakable role in the recruitment process, at least in the northeast region.

In other words, if one wishes to deal effectively with continental shelf regimes (ecosystems), it is necessary to shed at least some of the terrestrial ecological paradigms. That was my "administrative tone" for many years, with MSY (maximum sustainable yield) a particular "bete noir," in that regard. The use and abuse of that "paradigm" has been dealt with adequately in recent years so we don't need to discuss it any further here. But it is easy to see why so many people, including fishery biologists, "intuitively" understood it in the way in which they did. It makes good sense from the standpoint of one familiar with terrestrial ecosystems, where density dependent interactions play a dominant role in determining population size.

We have made significant progress in setting aside the terrestrial paradigms in relation to marine ecology. Enough long term data has accumulated to begin the effective dissection of marine regimes in the Northeast and elsewhere as well. Now we can really begin to specify some badly needed basic research activity. It should be emphasized, however, that the need has not abated to maintain and further develop necessary long term data bases.

Our ability to monitor the oceans has changed dramatically in recent years. The Coastal Zone Color Scanner (CZCS) and the satellite thermal data (Tyros and NOAA) have made it possible, cloud cover not withstanding, to see pigment and temperature at the sea surface in great detail, as well as observe the scaling of phenomena such as fronts and upwelling. Most unfortunately CZCS will apparently not be with us much longer but the polar orbiting thermal sensors will continue unabated. Such satellite data will ease the logistics associated with many types of vessel operations, particularly those involved in monitoring the physical environment. It should be noted in passing that the lack of a color scanner will be sorely missed. Much of the information that will be useful to biological oceanographers is not that data directly measured by any particular sensor, but knowledge derived from two or more different bodies of data, appropriately interacted to produce a very different "parameter," as recently demonstrated by Tvirbutas and McPherson (1984).

For a very long time now biological and fishery scientists have examined the relationship between marine populations and climate. The list of authors is long. In the northeast they include Chase (1955), Sissenwine (1974), Dow (1977), Sutcliffe et al. (1977), and many more. There is a general progression seen in all these studies here and elsewhere, from those based simply on long term atmospheric climate trends to running averages of two to four years past of air and ocean temperatures, and lately a few dealing with rather specific ocean climatic data collected at the time of spawning.

It is apparent that there is a hierarchial array of atmospheric influences on currents and temperatures, beginning with major atmospheric events of a long term, global character—the trade winds for example, to mesoscale events such as the persistence and strength of particular air masses (e.g. the Bermuda High and Icelandic Low). This topic has received a great deal of attention in recent years in the North Atlantic (cf. Lamb and Ratcliffe, 1969) and elsewhere.

The earlier works on fishes mentioned above demonstrated statistically significant, although unexplained, relationships. They dealt with water temperatures and atmospheric phenomena, usually those that occurred prior to the fish population events of interest. In this respect it is heartening that the past appears to foreshadow the future as much as it seems to in such matters. We still have a long way to go before we can predict the atmospheric weather itself more than just a few months ahead, if even that much.

The Northeast Fisheries Center recently processed the available physical oceanographic data for the region from Cape Hatteras to Nova Scotia. The temperature data is, as elsewhere, the bulk of the data. For the period 1941 to 1980 the region has experienced some significant shifts in annual temperature as shown in

Figure 16.1. The earlier period, that prior to 1945, is less well represented in the data base, but it is at least consistent with what follows. A marked cooling trend occurred between 1958 and 1968, culminating sharply. The surface layer (0 to 100 m) temperature differences between years largely reflect the atmospheric temperatures of the winter months (December through March).

The marked shift in average temperatures shown was accompanied by some dramatic changes in fish populations, one of which seemed unrelated to fishing activities--the upsurge in mackerel, and the other, the virtual demise of the haddock population, to overzealous fishing. There were other changes as well, but the data available for the other species was less than adequate even for speculating on the causes for the changes.

Recently I had the opportunity to examine the role of water temperature as it related to haddock recruitment. The haddock population in the northeast is subdivided into a number of reasonably discrete stocks with that on Georges Bank being a particularly large and consistently productive one. For many years it supported a substantial fishery, capable of providing, on the average, from 30 to 50 thousand metric tons a year. This particular stock spawns on the eastern end of Georges Bank in the early spring. The eggs and larvae tend to be retained on the bank by the current system, which includes a "gyre" on the shallower (60 m or less) portion of the bank. The "gyre" develops in part as a consequence of tidal action and in part because of the current regime that is created by density differences between the bank and surrounding water masses.

The area in which the haddock spawn on eastern Georges Bank is a mixing bowl and the place where the Middle Atlantic "cold pool" originates. Included in this mix is upwelled water from the Northeast Channel, water from the gyre on the bank, and water both from the surface and intermediate layers of the Gulf of Maine. As yet there is little information on the mix seasonally, but it tends to be relatively unvarying in characteristics, including temperature during the warmer months of the year.

The eggs and larvae from this spawning area will tend westward, with some being entrained by the gyre. Under usual circumstances they will grow large enough to maintain themselves in the Georges Bank area before they are swept further west and south into the southern New England region. In a few instances, notably when the year classes have been unusually large, juveniles have been taken in the fall as far west as the Hudson Canyon. Among the factors that play a role in

292

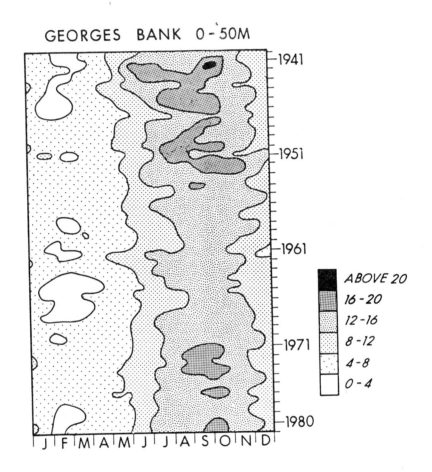

GEORGES BANK 0-50M

Figure 16.1. The seasonal water temperatures observed
in the surface layers (0 to 50 m) in the Georges Bank
area, 1941 to 1980. Data from NODC files and processed
at the University of Massachusetts Computer Center.
The data for the earlier period, especially that collected during the war years, are suspect.

determining their fate are: (1) the temperature of water, (2) the speed of the along-shelf current at the southern periphery of the gyre, (3) the amount of water recirculating at the western end of the gyre, and (4) the number of eggs and larvae lost at the outer margin.

The details of the study of haddock recruitment and water temperatures will not be presented here. They will be published shortly (Edwards, 1984).

Water temperatures for the cold period--January, February, and March--accounted for over 40% of the variability of haddock recruitment. Water temperatures for the following cold period, when included in the analysis, accounted for over 60% of the variability. Temperatures at other times of the year seemed to have little significance. So now we have a suggestive relationship based on temperature just prior to and at the time of haddock spawning and one year later. That seems to imply that it is not water temperature per se, but rather that the nature of the current regime--in this instance we are probably talking about the gyre on the bank, its size and constancy, and the related fronts--may have more to do with the survival of eggs and larvae than temperature itself.

Figure 16.2 is a plot of the temperatures for the two periods, 1946 to 1980, and the log of the recruitment index. The recruitment index is calculated by dividing the number of recruits (based on estimates of 1-yr-old fish) by the numbers of spawning fish (ages 3 and over).

Under most circumstances temperature is but a symptom of events in the current regime and not directly descriptive of them. As demonstrated by the work of Lasker (1978), and his colleagues, the stability of the water column is a critical factor for some species, since it determines the relative density of food organisms available to fish larvae. Again water temperature is not descriptive of that factor, although it may be symptomatic of it under certain circumstances.

From my point of view the present priorities for advancing the study of very large marine ecosystems--regimes--are:

(1) The need for greater emphasis on continental shelf current and water column "structuring" regimes and the driving forces involved. The physical oceanographer has much to do in assisting the fisheries biologist, in the identification of, and study of the variability of the environmental phenomena that enhance or disrupt aggregation and survival of marine populations, and

(2) The need to maintain and improve monitoring capabilities, both for the physical factors already recognized as relevant and new factors as they are

294

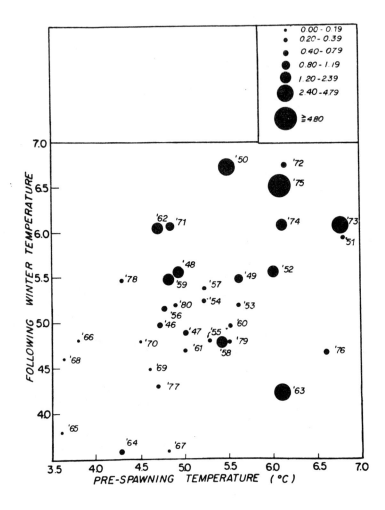

Figure 16.2. The recruitment of the Georges Bank haddock stock as it related to water temperature at the time of aggregation for spawning of each year class (January, February and March) and for the same time period the following year.

identified by the oceanographers as influencing significant population changes in the biota.

The first priority can best be achieved in the
basic research community, and I am sure that the new
remote sensing technology will play a prime role in
this effort. It should be noted, however, that I am
not aware that much of this work is presently going on.

The second priority is of prime concern to the
mission oriented agencies, and will obviously be aided
and abetted by the new ocean buoy technology.

To go back to the earlier theme, the organisms of
ocean ecosystems do not have the vegetational go-
betweens of terrestrial systems. The near term pay-off
would appear to be a closer examination of the physical
dynamics that are associated with spawning and early
life history, followed by a detailed examination of the
interplay between these planktonic species, plant and
animal, as they are influenced by and respond to these
physical phenomena.

And lastly, there remains much work to be done on
the genetics of fish populations. Not enough has been
done on "r" strategists of the magnitude represented by
many continental shelf species of fishes. It is specu-
lative on my part but I would suggest that the combination of environmental change such as we have seen in
the northeast region plus excessive fishing has geneti-
cally "bottlenecked" some stocks and may be the explanation for some of the prolonged periods of reestab-
lishment of stocks that have been observed. Some
people suggest that fish stocks are ephemeral. Perhaps
so, but not any more so than terrestrial populations.
To my mind it is simply a matter of the different time
scale of events that influence comparable genetic
systems.

A note on the occasional use of the term "regime"
instead of ecosystem in my comments. One dictionary
definition of regime is "a system that is continuously
or progressively changing in a regular and definable
manner," as for example the seasonal cycle of rain-
fall. Continental shelf environments change like this,
of course. Continental shelf ecosystems, however, are
not characterized by the same aspect of seasonal and
annual stability provided for the biota associated with
the fixed and physically large vegetational intermedi-
aries of terrestrial ecosystems. While we can all
think of exceptions, among them coral reef communities
and benthic infauna, by and large the marine ecosystems
function very differently from their terrestrial coun-
terparts, in that the biota is directly responsive to
the physical environment.

In conclusion, everyone appreciates that salmon
and other anadromous species are adept at finding their
way back home. Iles and Sinclair (1982) and others

have shown that herring are but one step removed from
anadromous species, and respond to phenomena associated
with tidally stirred areas near shore. Evidence is
mounting that other species similarly aggregate for
spawning and other activities to phenomena such as
frontal areas, regions of upwelling and so forth.
Though their spawning strategies may differ, each
marine species seems to carry out its activities by
responding to abiotic, often dynamic, aspects of their
environment. Marine ecosystems are bounded by such
frontal areas, and, like those on land when one considers individual species, they are difficult to
define. But it becomes easier to deal with the situation if one considers that the physical environment
plays much the same role in the life history of fishes
as does the tree in the life history of squirrels.
Since there appears to be considerable predictability
in the shaping of these features, their study ranks
high on my list of priorities.

COMMENT

    I particularly liked the "systematization" of
phenomena scaling and the stress on interregional comparisons presented in the paper by Bakun (this volume).
Clearly there are a multitude of species involved
around the world and a comparable multitude of species
adaptations. Interregional comparisons can shed a lot
of light quickly on the scope of systems to which
organisms adapt and maintain self-supporting populations. I hope as well that a great deal of attention
will be paid in the future to sorting out the significance of the different time scales that the atmosphere
has on influencing ocean climate. Many of the relationships noted to date appear to be related more to
longer term influences than to short-term environmental
events. Since prediction is the name of the game in
fisheries research, the matter is understandably
important. Until short-term events in the atmosphere
can be predicted far better than presently, if at all,
the longer term influences at least offer some
considerable predictive capability.

REFERENCES

Chase, J. 1955. Winds and temperatures in relation to
    the brood strength of Georges Bank haddock. J.
    Cons. Cons. int. Explor. Mer 21(5):1216-1230.
Dow, R. L. 1977. Effects of climatic cycles on the
    relative abundance and availability of commercial
    marine and estuarine species. J. Cons. Cons. int.
    Explor. Mer 37(3):274-280.

Edwards, R. L. 1976. Middle Atlantic fisheries: Recent changes in populations and outlook. Spec. Symposium 2, Am. Soc. Limnol. Oceanogr. pp. 302-311.

Edwards, R. L. 1984. Environmental issues affecting fisheries. Paper presented at Eighth Annual Seminar of the Center for Oceans Law and Policy. Cancun, Mexico, January 11 to 14, 1984.

Iles, T. D., and Sinclair, M. 1982. Atlantic herring: Stock discreteness and abundance. Science 215(4533):627-633.

Lamb, H. H., and Ratcliffe, A. S. 1969. On the magnitude of climatic anomalies in the oceans and some related observations of atmospheric circulation behaviour. Rapp. P.-v. Réun. Cons. int. Explor. Mer 162:120-132.

Lasker, R. 1978. The relation between oceanographic conditions and larval anchovy food in the California current regions: Identification of factors contributing recruitment failure. Rapp. P.-v. Réun. Cons. int. Explor. Mer 173:212-230.

Sissenwine, M. P. 1974. Variability in recruitment and equilibrium catch of the southern New England yellowtail flounder fishery. J. Cons. Cons. int. Explor. Mer. 36(1):15-26.

Sutcliffe, W. H., Jr., Drinkwater, K., and Muir, B. S. 1977. Correlations of fish catch and environmental factors in the Gulf of Maine. J. Fish. Res. Board Can. 34:19-29.

Tvirbutas, A., and McPherson, C. 1984. Application of image processing techniques to marine fisheries. Paper prepared for presentation to Oceans "84" meeting, Washington, D.C. Sept. 9-12, 1984. Laboratory Reference No. CSDL-P-1921.

17. Large Marine Ecosystems and the Future of Ocean Studies: A Perspective

The first symposium on the Variability and Management of Large Marine Ecosystems (LMEs) is an attempt to come to terms with how best to measure changes in the oceans and preserve their enormous capacities for: (1) mediating climate and weather, (2) serving as a reservoir of oil, gas, polymetalic sulphides and other minerals, and (3) sustaining production of fisheries resources.

It is appropriate to focus on the importance of the oceans now. In fact, 1984-1985 has been designated as the Year of the Ocean in the United States. The objective of the Year of the Ocean is simply to foster a better understanding of the importance of ocean resources by strengthening the dialogue among those who use the sea to foster both its productivity and its health. We are now on the threshold of a new era. The future will require that we manage on an ecosystems basis. We can no longer view events in the marine environment as isolated. In a single decade, from 1970 to 1980, we have initiated the transition from resource exploration to resource management with the extension of national sovereignties of maritime states around the globe to 200 miles and in some instances, beyond. We need to focus on the significance of these events and their effect on the future of the oceans. Before addressing my remarks to the future, I should like to set them into a perspective.

As the largest feature on the globe, the ocean touches us all. We have eaten its food, bathed in its waters, sailed on its surface, and just plain watched its magnificence at the seashore. These are the pleasant experiences. There is, however, in the context of spaceage oceanography, a Star Wars perspective that addresses the "darker side of the force." We are using the oceans as a depository for wastes, including municipal sludges, radionuclides, petrogenic hydrocarbons, organochlorine pesticides, and other hazardous substances. We have overexploited fish stocks. We have depleted stocks of whales and other marine mammals. If unchecked, we may cause irreparable damage to the natural balance of ocean ecosystems. What then of the oceans' future? On balance, I am optimistic. I believe the future will be a period of a holding pattern on damage and a more enlightened approach to the conservation and management of both renewable and non-renewable resources.

New sources of oil, gas, and minerals are in the process of discovery and production off our shores and in other parts of the globe. Within the coastal waters of the United States these exploratory and production operations are being monitored to protect the marine ecosystem. Institutions, both nationally and internationally, are now in the throes of attempting to implement a new holistic approach to the management of renewable and non-renewable marine resources through the concept of ecosystem management.

Advances made in support of large marine ecosystem management are for the most part regional. They are the result of a good deal of research and monitoring activity underway within the recently-designated extended jurisdictions, particularly in the North Atlantic and including the Exclusive Economic Zone of the United States, encompassing approximately 2.2 million square miles of ocean within 200 miles of the U.S. coast. Within this zone, NOAA is in partnership with academic institutions through Sea Grant, other Federal Agencies, and with the private sector community to move ahead in optimizing yields from the oceans' resources consistent with a multi-use policy.

In his fine book, Analysis of Marine Ecosystems, published in 1981, Alan Longhurst unknowingly set the perspective for this symposium. As a scientist heading a large marine research organization, he is acutely aware of the increasing demands for information in support of enlightened ocean resource management. But as most of us involved in making management decisions recognize, our position on the information curve is well behind the demands made of us by the producers and users of ocean resources. Longhurst points out in his preface that quantitative ecology relating to environmental impact assessment is less than 10-yr old. He states that to measure progress from the observational, qualitative ecology of the recent past, toward the predictive, quantitative ecology which is already demanded, but not yet available, it is important to sit back and examine, with a synoptic view, the present status of marine ecology. He goes on to state that "only by measuring our rate of progress can we hope to retain credibility for the view that continued

support should be given to the long-term aspects of
ecological research in the coming decades, so as to
avoid a future in environmental management constrained
by scientific undercapitalization." In retrospect,
Longhurst is correct. We are, indeed, collectively
engaged in an intellectual pursuit of the appropriate
strategies for measuring marine ecosystem variability.
We, in NOAA, recognize that this is not a short-term
problem and that governments are obligated to invest in
long-term studies if we are to ensure long-term
benefits.

NOAA AND LMEs

NOAA is actively conducting research and monitoring of large marine ecosystems within the EEZ. The
major ecosystems are defined by unique bathymetry,
hydrography, productivity, and trophically-linked
population structures. They include the Northeast
Atlantic shelf, Southeast Atlantic shelf, Gulf of
Mexico, California Current, Gulf of Alaska, Eastern
Bering Sea, and Insular Pacific.
There is an increasing public awareness of the
potential damage to marine ecosystems from the impacts
of pollution and overexploitation. In the United
States legislative authorities have mandated responsibilities to the federal government to deal with these
issues. However, the mandates do not reside within a
single federal agency. For example, responsibility for
protecting the public from health hazards emanating
from ocean pollution resides with the Public Health
Service, the Food and Drug Administration, and the
Environmental Protection Agency. The Corps of Engineers is responsible for protecting the seashore and
removing hazards to navigation; mapping is done at the
National Ocean Survey, protection of marine mammals is
a shared responsibility with the Marine Mammal Commission and NMFS, and marine birds are the responsibility
of the Fish and Wildlife Service. The assessment and
management of marine fish, marine mammals and their
habitat is a prime responsibility of the National
Marine Fisheries Service, which also has jurisdiction
over endangered species. Offshore gas, oil, and
mineral development is regulated by the Department of
Interior and energy-related marine impacts are the
concern of the Department of Energy. The listing is
impressive and does not include the responsibilities of
the Coast Guard, the Navy, and other service organizations.
Virtually all of the living resources and habitat
assessment and monitoring within the LMEs of the United
States are the responsibility of NOAA and its mainline

components, including the National Ocean Service, the
National Marine Fisheries Service, and the Environmental Research Laboratories. NOAA has recently established an Estuarine Coordination Office that will serve
as a focus for activities in U.S. estuaries. NOAA is
adopting a marine environmental quality program that
identifies areas of concern including nutrient overen-
richment and habitat alteration. Further, the National
Ocean Service has undertaken a pollution monitoring
program in marine, coastal, and estuarine waters. NOAA
operates ships, laboratories, and programs that are
serving as the core of a national effort designed to
monitor changes within marine ecosystems of the EEZ. I
am making greater use of Memoranda of Understanding
among federal agencies to overcome institutional constraints within the federal government to holistic management of LMEs. We are making progress. Each year we
produce assessments of the major fishery stocks and
their environments. We also produce the nation's
National Marine Pollution Program Summary, and we
conduct joint studies with other agencies on problems
of mutual interest. NOAA, as a proponent of more
interagency activity, is exploring means for fashioning
a national policy that will lead to total ecosystem
management within the EEZ. It is an important goal
that we are vigorously pursuing. I am convinced that
the federal government is now adequately funded to deal
with the ocean problems in the long term. But better
planning and coordinating of ocean studies, particularly within the EEZ, is clearly in order, and I intend
to pursue this goal vigorously.

As a relatively new federal agency involved in
studies of the atmosphere and oceans, we in NOAA find
ourselves with increasing demands for providing
services. Among these services are forecasts of ocean
weather and climate, predictions of fish stock abundance, reports on bathymetry, water mass structure,
circulation, and environmental monitoring measurements
at the sites of marine mineral explorations. We are
also concerned with adequately measuring and addressing
the growing problem of coastal pollution and habitat
loss. These, of course, are not interests unique to
the United States. We are aware of the growing concerns in other maritime nations of the presence of
detectable levels of organochlorines, trace metals, and
petrogenic hydrocarbons in coastal waters. I am
pleased to report that NOAA, through the National
Marine Fisheries Service, will, during 1985, participate in a monitoring program sponsored by the International Council for the Exploration of the Sea to
evaluate the health of the marine environment over the
entire North Atlantic area.

LME MONITORING

NOAA is actively conducting directed research and monitoring of Large Marine Ecosystems within the EEZ. Ship operations during 1985 include extensive bathymetric measurement, fishery assessments, marine mammal assessments, and environmental assessments of the continental shelves, off the northwest coast, Alaska, the Gulf of Mexico, as well as the east coast. In addition, the ocean assessment program will extend into the Central Pacific for explorations of sea mounts. This effort will be supported by 21 ocean-going vessels operating over a total of 4,251 sea days. We are committed to the large marine ecosystems management concept. We have in place the laboratories, scientists, ships, planes, and thermal microwave and ocean pigment satellites that are being used in addressing the population and environmental issues of concern.

OCEAN SERVICE CENTERS

To ensure that information from this effort is readily available, I have proposed the establishment of Ocean Science Centers to better serve the ocean community and public. The Centers will provide, on a timely basis, information available from NOAA on, for example, the status of ocean weather, fisheries, bathymetry, currents, and pollution assessment for use in tactical or longer-term purposes. One of the Centers is already in operation providing Alaskan users with information on ocean products within the EEZ in the Gulf of Alaska and Eastern Bering Sea.

NEW APPROACHES TO FISHERIES ECOSYSTEM MANAGEMENT

While we may have the dubious distinction of creating yet another acronym in our use of LMEs for large marine ecosystems, the concept was very much a topic of discussion in Europe where events were moving forward toward ecosystem management during the mid-1970s, predicated on shifts in fish stock abundance. The new attitudes for coming to grips with multispecies overfishing and options for ecosystem management were discussed at length in meetings of the International Commission for Northwest Atlantic Fisheries (ICNAF) during the early 1970s. We had many fruitful discussions with our European colleagues on this issue. The concept of managing fishery resources within the spatial limits of a large marine ecosystem was advocated by ICNAF. In 1972, ICNAF established the first two-tiered system for managing the total finfish biomass within the Northwest Atlantic ecosystem. Catches were capped at a level of one million metric tons annually for all species combined, and total allowable catch limits were allocated for each of the important species comprising the biomass.

Some progress was made in reducing heavy fishing mortality under ICNAF. However, the recovery process for bringing depleted stocks to former abundance levels was slow. Three years later, in 1976, the ICNAF model was replaced with the extension of exclusive fisheries jurisdiction to 200 miles off the coast of the United States, with Federal legislation establishing this area as the Fishery Conservation Zone.

Although it is an anathema to legislate science, the language used in establishing the FCZ served as an important step toward ecosystem management. The legislation enacted in 1976 called for fishery conservation and management concerning the interdependence of fisheries or stocks of fish, the impact of pollution on fish, the impact of wetland and estuarine degradation, and other matters bearing upon the abundance and availability of fish.

During this same time period of the mid-1970s, our European colleagues were making substantial progress in identifying new approaches to fisheries ecosystem management. In 1975, during a symposium concerned with changes in the North Sea fish stocks, the convenor, Dr. Gotthilf Hempel of the Federal Republic of Germany and former president of ICES, focused attention on the importance of interactions between the various fish stocks in the North Sea. His observations were based on the pioneering fishery ecosystem models by the Danish scientists, Drs. Andersen and Ursin. They hypothesized that predation and food composition were density-dependent functions of both predator and prey abundance. The formerly abundant stocks of herring and mackerel presumably took major amounts of various kinds of fish larvae, and in turn, they pointed out that herring and mackerel were preyed upon by large gadoids, particularly cod. Therefore it was hypothesized that the pelagic stocks would benefit from a reduction in the average size of gadoid predators. In fact, the total yield of fish from the North Sea would approximately double, based on the simulation results. That is an increase from about 3 million tons to about 6 million tons, although the value of the catch would probably increase by a smaller amount because of the smaller size of the fish. Hempel appropriately credited Andersen and Ursin with advocating a new policy of fishery management for the North Sea based on a multispecies approach that advocated higher fishing effort on large predatory species.

The concept of multispecies management has been pursued by the NMFS and NOAA since the early 1970s, and the effects of predator-prey interactions and competition are being considered in conjunction with changes in fishing effort to increase fish yields. Some of the pertinent model simulations are being done on resources just off the northeast coast on Georges Bank stocks. Georges Bank has the highest yield of fish per unit area of any north temperate marine ecosystem. This yield represents only a small fraction of the fish production on the Bank. About 85% to 90% of the annual fish production on Georges Bank is consumed by predators, including fish, mammals, and birds. Here we have the stuff of real progress in our approach to the conservation and management of fishery resources. Conceptually, we could increase the total fish yield by perhaps a factor of two by selectively cropping the larger fish-eating species in the Georges Bank ecosystem. While this is conceptually possible, there remain significant uncertainties in the biological, economic, and technological interactions in a multispecies fishery that need to be carefully evaluated before encouraging selective predation on a large scale.

TRAINING ETHIC FOR EEZ RESPONSIBILITIES

I should like now to turn my attention to another issue, and that is, can large scientific research and monitoring operations be successful? I think the answer is yes. However, we need to do a better job in the coordination and integration of the multidisciplinary science and technology required to ensure success. We have decentralized our research and monitoring operations to allow for improved decision making at the local operating levels in NOAA. Our decentralization of operations has the advantage of tight, on-sight management control. But we are going to have to work harder at coordination and integration at the program level and with appropriate data syntheses at the working scientist level. This is not an insignificant challenge. In all our EEZ operations we are developing joint projects with our academic colleagues. But you must keep in mind the nature of graduate training, by tradition, emphasizes individual creativity and initiative. Both are worthy virtues leading to the development of academic superstars. But if we are to deal effectively with the management of marine ecosystems, we need more than the substantial logistical support for conducting measurements of the oceans and their populations. We need to foster a new ethic that stresses a multidisciplinary team approach to ocean science. Each of our creative scientists will need to give up a bit of independence for the greater good.
We are making progress in dealing with this intellectual problem. A growing number of our NOAA

scientists are serving as guest lecturers and adjunct faculty and they are spreading the message. As the concept filters into our university training programs, our graduating scientists should become better able to cope in the future with the real-world adjustments needed for effectively doing team-science.

In my three years with NOAA, I have come to appreciate and admire the willingness and capabilities of our scientists, technicians, and supporting staff for getting on with the job. I'm convinced that we have made significant progress and I fully expect that NOAA, in the next decade, will be in the forefront of scientific organizations supporting the effective management of large marine ecosystems. We already have initiated programs for monitoring biological and environmental variability within the EEZ in the operations of NOAA's main line components--National Weather Service, National Ocean Service, National Marine Fisheries Service, Office of Oceanic and Atmospheric Research, and the National Satellite Service. And the coordination links across NOAAs Main Line Components (MLCs) are being strengthened. NOAA's expertise allows it to serve as the scientific spokesman for marine issues of national concern.

INTERNATIONAL EFFORT

Now, what of the apparent contradictions of extended coastal jurisdiction on the one hand and the need for even greater international cooperation in ocean research on the other? Two events of the past year, El Niño and the possibility of a partial krill recruitment failure in the Antarctic, illustrate the importance of the need for even greater international cooperation. The flip-flop of atmospheric pressures in the Pacific apparently fueled the massive El Niño. This is an event resulting in hundreds of millions of dollars in damage in both hemispheres including losses of millions of tons of anchovy catch off Peru, drought in Australia, humid air over South America causing intensive precipitation and flooding, and heavy rains and flooding on the U.S. west coast. NOAA is playing an active role in attempting to better understand the events leading to the atmospheric oscillations that triggered El Niño. This effort is being pursued within the international ocean community.

The krill situation observed in the Antarctic this year may have, in fact, been related to El Niño. Our scientists have recently returned from the Antarctic where they report that the temperatures are about 2°C above normal and biomass of krill is apparently 30 times less than it was last year. In the light of this new information, the estimates of krill biomass which some scientists claim to be 10 times the 70 million metric tons of present world catch of marine fish, may now be in need of revision based on the low levels of krill observed by the U.S. team of NSF and NOAA scientists. As a delegate to the International Whaling Commission, I am very much concerned by the apparent decline in krill abundance in the Antarctic marine ecosystem and the consequences that a decline may have on dependent populations of whales, birds, and fish, but I am also encouraged that the new Commission for the Conservation of Antarctic Marine Living Resources (CCAMLR) is attacking the whale-krill-ecosystem linkage problem. The Commission is the first to tackle the management of a total marine ecosystem and we in NOAA are committed to its success. We are preparing a plan for greater NOAA participation in CCAMLR because we believe that it is an important milestone in the effort to effectively manage, for optimum yield, multispecies fisheries operating at different trophic levels.

We in NOAA intend to remain active in international marine affairs. We cannot allow the extended jurisdictional regionalization of the oceans to become a divisive force. During this Year of the Ocean, I'm pleased to report to you that NOAA is moving forward in LME research nationally within the EEZ, regionally in joint programs with other countries in the Pacific, in our participation in EPOCS and NORPAC, and globally in studies of sea-floor spreading and plate tectonics.

As new discoveries are made with greater frequency now that research and assessment efforts are increasing within the territorial seas, we will need to work even harder exchanging scientists, conducting joint research operations, and convening symposia and workshops to nurture international cooperation. I look forward to greater international participation in the open-ocean waters. We have an unprecedented opportunity for demonstrating to the world the great benefits in joint international studies of the oceans. I look to the Antarctic as the model for the future. The fourteen nations, including Poland, the Soviet Union, and German Democratic Republic, that have joined the Commission for the Conservation of Antarctic Marine Living Resources are working toward the implementation of total ecosystem management. We stand ready to work with them and other nations committed to achieving success in restoring the great whales while ensuring optimal utilization of both renewable and nonrenewable Antarctic resources.

CONCLUSION

I remain optimistic about the future of the oceans. For each of the great discoveries we may

experience temporary setbacks, but these can best be
handled within the context of progress. We can look to
the future and realistically move forward, recognizing
that the challenge is great and the way difficult, but
keeping in mind Theodore Roosevelt's conviction that:

> Far better it is to dare mighty things, to
> win glorious triumphs, even though checkered
> by failure, than to take rank with those poor
> spirits who neither enjoy much nor suffer
> much, because they live in the great twilight
> that knows neither victory nor defeat.

I share in the excitement that is generated when
new concepts "daring mighty things" are discussed and
debated. Marine resources management is headed in the
right direction. We must step back often enough from
the mundane demands of daily activity to consider the
really big question of how best to measure ecosystem
variability. The solutions discussed in this volume
underscore my optimism for the future of the oceans.

Index

BIOMASS. See Biological
    Investigations of
    Marine Antarctic
    Systems and Stocks
Biomass, 10, 15, 66, 71,
    75, 78, 94
    benthic, 4, 26
    changes in, 5
Birds, 5
Bluefish, Pomatomus
    saltatrix, 71, 226
Blue-green algae, 24-25
Bocaccio, Sebastes sp.,
    42, 44
Bongo net, 211
    plankton surveys, 71,
    207
Bonito, 99
    Pacific, 42
Bottom trawl surveys, 71
Bream, 4, 26
Bulletins Statistiques,
    ICES, 146, 148. See
    International Coun-
    cil for the Explora-
    tion of the Sea

CalCOFI. See California
    Cooperative Oceanic
    Fisheries Investiga-
    tions
California Cooperative
    Oceanic Fisheries
    Investigations
    (CalCOFI), 35, 42,
    46, 207-208
    See also model,
    CalCOFI
California Current, 4, 5,
    7, 94, 99, 203, 239,
    301
California Fish and Game
    Commission, 50
California sea lions,
    47. See also
    Pinnipeds
Canary Current LME, 5
Capelin, Mallotus
    villosus, 111
Carbon-14, surveys, 71
    values, 214
Carrying capacity, 11
Catch, 270, 273, 279

Catch per unit effort
    (CPUE), 120, 146
CCAMLR. See Convention
    for the Conservation
    of Antarctic Marine
    Living Resources
Central Pacific, 303
Chendytes lawi, 38
Chilipepper, Sebastes
    sp., 42
Chlorophyll, 6, 175-176,
    178, 182, 185, 188,
    191, 195, 198-199,
    211
    Chlorophyll a, 175-176
Cladophora, 26
Climatic variations, 19-
    23
Clupeid fish, 127
Clupeid species, 12
Coastal Zone Color
    Scanner (CZCS), 289-
    290
Coastal zone sediments,
    25
Cod, 4, 6, 27, 62, 71-80,
    127-128, 130, 149,
    152-154, 157, 162,
    165, 168-170, 225,
    304
    Baltic, 265
    Gadus macrocephalus,
    127, 130
    G. morhua, 208, 211
Cold pool, 60
    Middle Atlantic, 291
Commission for the
    Conservation of
    Antarctic Marine
    Living Resources,
    230, 285, 307
Common Fisheries Policy,
    265
Common property, 263-264,
    271, 277-278
Community abundance, 15
Community change, 15
Competition, 14, 40
Competitive replacement,
    40
Conductivity, 175
Conservation zones, ex-
    tended fishing, 247